Excessive Inequality and Socio-Economic Progress

The growing inequality in the global economy across the planet is reaching unprecedented levels. This book seeks to develop frameworks for the assessment of excessive inequality and its impact on social-economic progress and sustainable development.

It begins by summarizing the theoretical approaches of economic inequality, its specificity, and questioning what economic inequality really is and how it progresses. Next, the book explores issues of methodology for addressing the growing excessive economic inequality. It then applies these concepts to examine inequality across a range of the European Union (EU) countries. A variety of factors are considered, such as the impact of economic inequality on socio-economic progress, when normal inequality turns into excessive inequality, and its impact on economic growth, quality of life, and the environmental sustainability across different groups.

Prof. Hab. Dr. Ona Gražina Rakauskienė is a professor at Mykolas Romeris University and Head of Quality Life Laboratory at Mykolas Romeris University.

Prof. Dr. Dalia Štreimikienė is a leading researcher at Mykolas Romeris University.

Dr. Lina Volodzkienė holds PhD of economics and is a lecturer at Mykolas Romeris University.

The Dynamics of Economic Space

This series aims to play a leading international role in the development, promulgation and dissemination of new ideas in economic geography. It has as its goal the development of a strong analytical perspective on the processes, problems and policies associated with the dynamics of local and regional economies as they are incorporated into the globalizing world economy. In recognition of the increasing complexity of the world economy, the Commission's interests include: industrial production; business, professional and financial services, and the broader service economy including e-business; corporations, corporate power, enterprise and entrepreneurship; the changing world of work and intensifying economic interconnectedness.

Rural–Urban Linkages for Sustainable Development
Edited by Armin Kratzer and Jutta Kister

Beyond Free Market
Social Inclusion and Globalization
Edited by Fayyaz Baqir and Sanni Yaya

Culture, Creativity and Economy
Collaborative practices, value creation and spaces of creativity
Edited by Brian J. Hracs, Taylor Brydges, Tina Haisch, Atle Hauge, Johan Jansson and Jenny Sjöholm

Social Protection and Informal Workers in Sub-Saharan Africa
Lived Realities and Associational Experiences from Tanzania and Kenya
Edited by Lone Riisgaard, Winnie Mitullah, and Nina Torm

Economies, Institutions, and Territories
Dissecting Nexuses in a Changing World
Edited by Luca Storti, Giulia Urso and Neil Reid

Excessive Inequality and Socio-Economic Progress
Ona Gražina Rakauskienė, Dalia Štreimikienė and Lina Volodzkienė

Excessive Inequality and Socio-Economic Progress

Ona Gražina Rakauskienė,
Dalia Štreimikienė, and Lina
Volodzkienė

LONDON AND NEW YORK

First published 2022
by Routledge
4 Park Square, Milton Park, Abingdon, Oxon OX14 4RN

and by Routledge
605 Third Avenue, New York, NY 10158

Routledge is an imprint of the Taylor & Francis Group, an informa business

British Library Cataloguing-in-Publication Data
A catalogue record for this book is available from the British Library

Library of Congress Cataloging-in-Publication Data
A catalog record has been requested for this book

ISBN: 978-1-032-23493-9 (hbk)
ISBN: 978-1-032-24378-8 (pbk)
ISBN: 978-1-003-27831-3 (ebk)

DOI: 10.4324/b22984

Typeset in Bembo Std
by KnowledgeWorks Global Ltd.

Contents

Figures

Tables

INTRODUCTION

The growing inequality in the global economy across the planet is reaching unprecedented level, especially in recent decades. Economic inequality, with its negative economic, social, and political consequences, is becoming one of the most pressing global challenges threatening the sustainability and socio-economic progress of the whole world. According to Nobel laureate J. E. Stiglitz (2012), all social, political, economic problems and their tragic consequences are embedded in inequality. The extent and depth of global problems linked to economic inequality is confirmed by these facts.

First, according to global surveys, 10% of all world's richest people dispose of more than half (52.10%) of world total income and 1% the super-rich account for as much as one-fifth (20.40%) of income. Meanwhile, half (50%) of the world's poorest people dispose only a tenth (9.7%) of total income (World Inequality Database, 2020).

Second, in the World Economic Forum Annual Meeting 2019 in Davos, a nonprofit organization fighting poverty and deprivation, Oxfam pointed out that 26 of the world's super rich own as much wealth as $3.8 billion of the world's poorest people ($1.4 trillion). In 2018, the wealth of the super-rich people increased by 12% ($900 billion), i.e., increased by $2.5 billion daily. Meanwhile, the financial situation of the poorest deteriorated; and during 2018, the value of their wealth decreased by 11% (Lawson et al., 2019).

Third, Lithuania's annual survey of the richest people (2019) shows that the 500 richest people in Lithuania own 19.56 billion EUR, and almost a quarter (24%) of this amount is concentrated in the hands of the 10 richest people in Lithuania. Meanwhile, in 2020 Lithuania's budget was 41% lower than the value of wealth managed by the richest persons – 11.53 billion EUR (Seimas of the Republic of Lithuania, 2019).

In contemporary theoretical research on economic inequality, two essential approaches can be distinguished. According to the first, the liberal approach, economic inequality is the result of the modernization of market economy and society. Economic inequality is fully justified as a result of a market economy. In this sense, inequality is based on the marginal productivity theory, arguing that higher incomes are associated with higher productivity and, consequently, a greater contribution to the well-being

DOI: 10.4324/b22984-1

of society. According to the second approach, economic inequality is not always justified, and it is argued that inequality is a problem of the economic system, especially when it escalates into excessive inequality, which hinders socio-economic progress.

Socio-economic progress is a multidimensional process whose essence is to guarantee long-term prosperity not only for present, but also for future generations. The integrity of economic growth, quality of life, and sustainability is one of the basic preconditions for ensuring socio-economic progress. The main challenge in this process is to find a balance between economic growth and quality of life, but at the same time to satisfy condition of sustainable use of resources (Jackson, 2012). Thus, the goal of socio-economic progress is not only to ensure the well-being of society in the present, but also to guarantee no worse opportunities for well-being of future generations, balancing economic, social, and environmental development in the present.

Recent world research shows that inequality has a negative impact on socioeconomic progress. Representatives of the neoliberal economic model argue that in a market economy, an increase in inequality is inevitable as long as the economy is growing and its level of development is still low. However, other numerous world theoretical and practical research shows that such attitudes are questionable. Back in 2006 in the World Bank's report "Justice and Development" (World Bank, 2006), studies conducted in various countries around the world show that excessive *economic inequality hampers economic growth.* Rising economic inequality significantly increases only the income of the rich and reduces the income of all the rest of the population, impoverishing the majority of the population. As a result, the economy becomes inefficient and reduces opportunities to invest and innovate.

Nobel Laureate J. E. Stiglitz in 2015 in his report "Rewriting the Rules" (2015), submitted to the Roosevelt Institute, states that economic inequality is not inevitable – it is not a social problem but an economic one. Inequality is not the price of economic growth, as liberals argue, but is the cause of slowing economic growth. According to J. E. Stiglitz, the higher is the inequality, the lower the economic growth. Inequality, according to J. E. Stiglitz, is a systemic problem, and a feature of the economic system.

Two main conclusions were made by J. E. Stiglitz in his report: firstly, inequality is the result of economic policy that favor the rich; second, the roots of this problem cannot be traced to any single fragmentary factor, such as corporate tax, health, or labor market reform. They consist of a set of factors. Therefore, the fight against inequality requires a systemic approach, a systematic solution in many areas – financial reform, corporate governance, tax policy, monopolies regulation, money, education, health, labor policies, etc.

Similar questions are raised by R. Reich, who argues that the cause of the global crisis is not an increase in public debt, a way of living and spending not according income, but the consequence of huge economic inequality, where GDP growth is based on unjustified increases in the income of the rich (Reich, 2010). R. Reich's research has shown that the credit boom of the last 30 years has been accompanied by a widening gap between the richest people in the United States and everyone else. Therefore, according to R. Reich, the crisis was the result of a growing inequality in income distribution.

Inequality also affects the well-being of people and quality of life. Research shows that the distribution of the created product, ignoring social justice, has a negative impact on the quality of life of the population – demographic processes, health, material situation of people, and reduced access to education and training (Stiglitz, 2015). In this context, it is not any economic inequality that is in itself conducive to development and competition, but that is extreme inequality that is treated as excessive inequality.

Inequality increases the insecurity and vulnerability of human and the state. According to J. E. Stiglitz, the World Bank research has shown that two issues are vital to people: insecurity and vulnerability. Vulnerability is perceived as a threat of a fall in living standards, which is of particular concern if there is a risk of a fall in living standards to deprivation level. The traditional one-sided pursuit of GDP growth by economists has led to a lack of focus on vulnerability.

One of the factors that "makes the biggest contribution" to increasing vulnerability is economic inequality, where the poor cannot overcome the difficulties of their life. "We need to look at inequality not as a moral problem but as an economic challenge, closely linked, first, to economic growth and, second, to increasing vulnerability" (Stiglitz, 2015).

Economic inequality in the current generation leads to unequal opportunities for future generations as well (Piketty, 2016). Viewed through the prism of human development, economic inequality not only denies the principles of social justice, but also reduces the opportunities for members of society to access education, parenting, health care, culture, quality housing, and the living environment (Rakauskienė et al., 2017). Economic inequality can lead to the exclusion of people not only in terms of income and wealth, but also in social life, preventing life satisfaction and reducing quality of life. Meanwhile, the adequacy of income and consumption, disposition, or management of property determines a person's material and moral security, self-confidence, self-esteem, and quality of life, and can promote a person's self-realization functions and creative potential necessary for social-economic progress.

The growth of the state economy does not yet guarantee the increase of social welfare, because at the same time there is a redistribution of national income, which leads to greater or lesser inequality of income, consumption, and wealth in the country due to different understanding of social justice and implemented socio-economic policy. This problem is emphasized by the European Commission (2020), which states that the overall growth of the Lithuanian economy hides growing socio-economic disparities in the regions, the critical situation in the field of poverty (which is almost 10%

higher than the EU average) and inequality. In other words, *the state fails to convert GDP growth into the well-being of the majority of people.*

Inequality in the distribution of resources traditionally remains one of the most pressing challenges in the Lithuanian economy. This problem arose after Lithuania entered the market as a consequence of privatization, but became especially acute after the global crisis of 2008, which has also hit our country hard. *Lithuania is the "anti-leader" in the EU in terms of inequality during 2016–2021.* In Lithuania, the income gap between rich and poor is widening, and poverty and deprivation remain a sensitive problem.

The growing polarization of society, when there is no middle class or it is very small, is a special reason for Lithuania's economic decline, which requires the state's attention, as this phenomenon promotes social tensions in society, social cataclysms – social threats, emigration, and may disrupt economic development.

Economic inequality is receiving increasing attention from researchers and politicians. The high level of income distribution that has developed in the current period is recognized as a phenomenon that contributes to the negative socio-economic consequences. The imperfection of redistributive levers in the economy has created the conditions for an economically unjustified concentration of resources, material goods, and wealth in the hands of a small group of the population – the richest one.

Today, we are seeing that the economic system is not population-oriented, so that growing inequality and its consequences for economic growth, well-being, and human development cannot be ignored. Excessive income distribution is one of the most pressing problems in the country, which, if left unaddressed, will make it difficult to implement long-term economic development strategies and medium-term programs in Lithuania.

Research on economic inequality is particularly relevant today and should be given adequate attention in each country. Appropriate public policies should also be developed to reduce economic inequality, minimize excessive inequality and tensions between members of society, as well as to ensure a dignified and quality life for each individual and the sustainable development of the country.

The impact of economic inequality on socio-economic progress is also not fully defined. Data on income and consumption inequality are the most commonly used to assess the state of economic inequality, while studies on material living conditions and wealth inequality are less common, and the assessment of excessive inequality is a relatively new and not well-addressed problem. Most research assesses the impact of economic inequality on national economic growth, but far fewer studies examine the impact of economic inequality on quality of life or overall socio-economic progress.

In this context, research on economic inequality is gaining relevance and paramount importance.

The following Nobel laureates paid great attention to economic inequality, factors, and causes, consequences, influence on socio-economic progress of countries: J. E. Stiglitz (2009, 2015, 2016, 2017), A. Deaton (2003, 2013, 2015), A. Sen (2009, 2015), and P. Krugman (2008, 2014, 2020). Also, other world-renowned scientists such s R. Reich (2010, 2012), T. Piketty (2014, 2015, 2016), E. Saez (2016, 2018), A. B. Atkinson (2011), A. Kaasa (2005), F. Bourguignon (2016, 2018), J. K. Galbraith (2009, 2016, 2018), etc., investigated economic inequality problems.

In the wake of the global financial crisis in 2008, many international organizations such as the World Bank, the Organization for Economic Co-operation and Development (OECD), the International Monetary Fund (IMF), United Nations (UN), European Union institutions: European Commission, Council of Europe, Eurostat, and others have also begun to pay particular attention to economic inequality.

The large scale of economic inequality in recent years raises the question of the extent to which such inequality can be justified and when economic inequality begins to have a negative impact on socio-economic progress. The authors of this monograph are just trying to answer this question.

Research on the impact of economic inequality on socio-economic progress provides conflicting evidence, arguments, and the different impact being defined. Research in this area can be summarized in three directions. Representatives of the first direction highlight the negative relationship between economic inequality and socio-economic progress (Alesina & Rodrik, 1994; Alesina & Perotti, 1994; Atkinson, Piketty, & Saez, 2011; Persson & Tabellini, 1994; Bourguignon, 2004; Rakauskienė, 2015, 2017, etc.). Proponents of the second direction argue that economic inequality has a positive impact on socio-economic progress (and GDP growth in particular) (Li & Zou, 1998; et al.). The third line of research suggests that there is a neutral relationship between economic inequality and socio-economic progress, that these impacts vary depending on circumstances and factors (Kuznets, 1955; Barro, 1999, 2000; Halter, Oechslin, & Zweimüller, 2013; Dominicis, Florax, & Groot, 2008; Hoeller, Jourmand, Pisu, & Bloch, 2012; Dabla-Noris et al., 2015; Grifell-Tatje, Lovell, & Turon, 2018). Unfortunately, the above-mentioned research directions do not reveal specific results; they are characterized by reasoning on a more theoretical, conceptual level.

Traditionally, economic inequality has been measured using income and consumption distribution methods, but increasingly, research on economic inequality analyzes the concentration of income in certain groups of society (mostly the rich). Although global research argues that the gap between the richest and the rest of society is widening, the issue of income concentration due to data availability and measurement complexity has not been fully explored. Special attention was paid to the problem of wealth concentration by T. Piketty (2006, 2014, 2015, 2016), R. Reich (2010), J. E. Stiglitz (2012, 2015, 2017), and others. Only a few scholars like J. E. Stiglitz (2015, 2017) have tried to distinguish the level of justified (normal) inequality from unjustified (excessive) inequality.

A number of researches on economic inequality (usually income distribution) have been conducted in Lithuania: R. Lazutka (2003, 2007, 2009, 2012, 2014), B. Gruževskis et al. (2012), A. Šileika (2009, 2010), O. G. Rakauskienė (2011, 2015, 2017), D. Skučienė (2008), Volodzkiene (2020), and others. However, there is still a lack of more in-depth research to determine how much of economic inequality can be justified – normal and how much of inequality can be treated as unjustified – excessive. It should also be emphasized that there is a lack of empirical research to make a reasonable assessment of the impact of economic inequality on socio-economic progress.

Currently, research on economic inequality pays special attention to the problem of the validity of economic inequality. *The research suggests that economic inequality can be normal and excessive.* A certain degree of inequality is justified, depending on education, qualifications, responsibility, as well as on the incentives created for innovation and entrepreneurship, on development, competition, and investment promotion. However, growing economic inequality becomes a problem when it restricts individuals' educational or professional choices, reduces a person's opportunities for self-realization, when his or her efforts are focused only on meeting the most basic needs. Excessive inequality is not just deep inequality, but from a certain level, it begins to hamper socio-economic progress. Excessive inequality negatively affects economic growth, well-being, and development of human resources and increases the vulnerability of population and the state.

In 2017, the European Commission has pointed out that "to some extent economic inequality can lead to investment in human capital, mobility and innovation", but "economic growth can be jeopardized if economic inequality becomes excessive". However, the scientific justification, namely when inequality becomes excessive, is lacking, and scholarly research on this topic is more of a theoretical and hypothetical nature.

According to the World Bank, economic inequality becomes excessive, ranging from 30% to 40% based on the Gini coefficient level. Excessive inequality is not just deep inequality (deep does not mean excessive), but one that, from a certain level, does not stimulate but stifles economic growth and has negative socio-economic consequences. In this sense, inequality is evolving from a social problem to a major impediment to economic growth and quality of life.

The negative social consequences of excessive inequality threaten, in particular, by creating a "poverty trap": people realize that they cannot break out from this trap, and there are no strong enough socio-economic incentives to do this, and people's motivation to take action is lost. This situation intensifies other negative phenomena: an increase in the number of suicides, an increase in the number of cardiovascular diseases, an increase in crime, etc.

In the light of the above arguments, a *scientific problem* arises which is caused by few main considerations: first, there is a lack of methods in the scientific literature to distinguish between normal inequality and excessive inequality; second, empirical research to objectively assess the impact of growing economic inequality on socio-economic progress is clearly insufficient. These circumstances make it possible to formulate a scientific problem and relevant research questions – *what is the real expression of economic inequality, to what extent it is justified, i.e., normal, and when it becomes unreasonable and unjustified, i.e., excessive and what is its impact on socio-economic progress.*

The study yielded relatively new and informative results presented in this monograph. *First,* based on theoretical assumptions and the multifaceted nature of economic inequality, after summarizing the methods of assessing economic inequality used in scientific literature and practice, *the concept of economic inequality based on normal and excessive inequality was developed and a model for assessing the impact of economic inequality on socio-economic progress was proposed.*

According to the authors, economic inequality is not neutral in terms of socio-economic progress; it affects economic growth, quality of life, and environmental sustainability of the countries, depending on the level of economic inequality achieved. Therefore, a differentiated approach to economic inequality, both normal and excessive, is needed. A certain level of economic inequality can be justified depending on education, qualifications, responsibility, as well as on the incentives created for innovation and entrepreneurship, on development, competition, and investment promotion. Meanwhile, inequality becomes excessive (unjustifiable), which, from a certain level, does not encourage, but on the contrary – hinder economic growth and causes negative socio-economic consequences.

Second, the study emphasizes a *differentiated approach to economic inequality – the model for assessing the impact of economic inequality on socio-economic progress in EU countries is based on normal and excessive inequality ratio.* Ways of separating these components of inequality and methods of assessing are proposed. Assumptions and hypotheses of this model are formulated. The uniqueness of results are based on the discovery of breaking points, when economic inequality turns from normal to excessive, and a marginal effect is identified, which quantifies how this excessive inequality negatively affects the socio-economic progress of the EU-28 (economic growth, quality of life, and environmental sustainability). Excessive inequality becomes apparent when the marginal effect of economic inequality changes from positive to negative; a further increase in economic inequality has a negative impact on socio-economic progress, slowing economic growth and reducing quality of life and causing danger to environmental sustainability.

Third, the impact of economic inequality on socio-economic progress in the group of EU-28 countries having different living standards is assessed. Empirical evidence suggests that economic inequality affects the socio-economic progress of most of the EU-28 member states. The high level of economic development achieved by some EU-28 countries does not yet guarantee protection against the negative impact of excessive inequality on the country's socio-economic progress – economic growth, quality of life, and environmental sustainability. The results of an empirical study confirm that excessive inequality has a negative impact on the socio-economic progress of both rich and less rich countries.

The current study shows that economic inequality does not affect socio-economic progress immediately and that the reaction of society and the economy is "delayed" and becomes apparent in the long run.

The negative effects of excessive inequality on socio-economic progress can be avoided through integrated systemic solutions aimed at reduction of the level of economic inequality. Empirical results of conducted study show that in order to avoid negative effects of excessive inequality on the socio – economic progress of the EU-28 countries, economic inequality measured by the Gini coefficient on disposable income, should not exceed 29–30%.

The monograph consists of four parts: (1) theoretical substantiation of economic inequality, (2) methodology for assessing excessive inequality, (3) assessing the impact of growing economic inequality on socio-economic progress, and (4) guidelines for reducing excessive inequality.

The first part of the book is devoted to summarizing the theoretical approaches of economic inequality, revealing the content and specificity of economic inequality, raising the problem of what economic inequality really is, whether it is justified, i.e., normal or unreasonable, i.e., excessive, developing the issue of wealth inequality, and to revealing a modern perception of socio-economic progress. The second part of the monograph deals with the issues of the methodology for addressing the growing excessive economic inequality. A concept of economic inequality based on the relationship between normal and excessive inequality is proposed; the methodological substantiation of the assessment of excessive inequality according to the clusters of EU countries is presented; the difficulties of the assessment of wealth inequality are revealed and the methods are proposed; the model of assessing the impact of economic inequality on socio-economic progress is formulated; the most important indicators of economic inequality and socio-economic progress are identified; the assumptions of the model are raised; and the econometric methods and procedures of impact assessment are presented.

In the third part of the monograph, according to the proposed methodology, income inequality in EU countries was empirically assessed by determining the income of the super-rich (which are not reflected in Eurostat). Clusters of EU countries are singled out according to normal and excessive inequality, as one of the first attempts to assess wealth inequality, the case of Lithuania is presented. The impact of economic inequality on socio-economic progress is assessed by identifying the breaking points of growing economic inequality, when normal inequality turns into excessive, and the impact of excessive inequality on economic growth, quality of life, and the environmental sustainability in different EU country groups (high living standard and lower living standards countries) is defined.

The authors would like to believe that the results presented in this monograph have not only theoretical but also practical significance. The developed methodology for assessing the impact of excessive inequality on socio-economic progress expands the field of inequality research methods and can be used by other scientists, statisticians, government officials, etc., to measure and assess national or regional economic inequality and its impact on socio-economic progress.

The model for assessing the impact of economic inequality on socio-economic progress presented in the monograph could be applied in formulating public policies that reduce economic inequality, minimize excessive inequality, reduce income inequality and tension among members of society, and promote dignity and quality of life as well as sustainable development of the country.

1 THEORETICAL BASIS OF ECONOMIC INEQUALITY

1.1 Development of economic inequality theoretical approaches

Economic inequality is one of the most pressing and debated topics in today's world. There are different approaches to the justification of inequality, its factors and causes, and the nature of the impact on socio–economic progress.

In the absence of a unified systemic approach to economic inequality, questions are often raised: (1) does the accumulation of private capital inevitably affect the increasing concentration of wealth in a very small (rich) group of people, as in the 19th century noted by K. Marx; (2) perhaps S. Kuznets in the beginning of 20th century was right in believing that the prevailing forces of economic growth, competition, and technological progress reduce economic inequality and promote harmony between classes in society; (3) is the distribution of income in the 18th and 19th centuries different from distributional inequalities in the modern world; (4) how the unequal distribution of income and wealth affects economic growth and quality of life (T. Piketty, 2014)?

The sources of economic inequality are demographic change, the need for food, and poverty. The issue of economic inequality became relevant at the end of 18th century and in the beginning of the 19th century, when in Great Britain and France underwent radical changes in society, influenced by extremely rapid demographic growth, the urbanization of the impoverished rural population, the need to feed themselves, and the ongoing industrial coup. These processes have affected the unequal distribution of wealth, changes in social structure, and political balance.

During this period, the widely known theory of T. Malthus emerged, which was published in 1798 and named "An Essay on the Principle of Population", indicating that the main threat to the stability of society is the surplus of the population. The findings of T. Malthus were based on the experience of France at a time when large numbers of poor rural people were flocking to the cities. T. Malthus, worried and trying to ensure that the same fate befell Great Britain, proposed the immediate cessation of any

DOI: 10.4324/b22984-2

supply or support to the poor and the limitation of its reproduction in order to prevent a population surplus leading to chaos and trouble.

Principle of unequal distribution. At the end of 18th century and in the beginning of the 19th century, radical economic and social changes took place. The insights of most economists at the time about long-term economic inequality and inequality in the distribution of society were gloomy and pessimistic. Some of the most famous and influential economists of the 19th century, D. Ricardo and K. Marx emphasized that a small social group (landowners and representatives of industrial capitalism) will inevitably absorb an ever-increasing share of income (Piketty, 2014).

In 1817, D. Ricardo in the book *On the Principles of Political Economy and Taxation* emphasized that landowners are absorbing an increasing share of total income, while the share of income of the rest of society is decreasing proportionally, thus disturbing the overall social balance. D. Ricardo considered that the only logical and politically acceptable solution was a steady increase in the taxation of land rental income (Piketty, 2014).

As history has shown, D. Ricardo's gloomy predictions were not correct: although land rent fees have remained high for some time, the value of arable land has also eventually declined as the importance of agriculture has diminished. D. Ricardo could not have foreseen the importance of future technological progress or industrial growth. Neither T. Malthus nor D. Ricardo could have imagined that the need for food would no longer be one of society's most pressing problems.

The insights of D. Ricardo on land prices have argued that the principle of unequal distribution and rapid price increases in a given area, can upset the balance around the world (Piketty, 2014). The latter principle can be applied to the economic inequality of recent decades. The increase in arable land prices mentioned in the D. Ricardo model can be compared with the real estate prices in the big cities of the modern world. As history has shown in 2008, the rapid rise in prices, especially in real estate, has upset the economic, social, and political balance, not only between different countries but also within each country.

Principle of unlimited capital accumulation. Fifty years later, economic and social trends have changed dramatically – the key issue was not whether farmers could feed a growing population or prices continue to rise, but the key question was how to explain the growth of industrial capitalism, which was flourishing at the time.

Like D. Ricardo, in 1867, K. Marx based his book *Capital* (volume I) on the analysis of the contradictions of the capitalist system. He sought to distance himself from bourgeois economists, who, based on A. Smith theory of the "invisible hand" and J. Say's "rule" argued that the output of production itself creates demand for itself; that the market is a system that can create and achieve a perfect balance on its own.

K. Marx used the D. Ricardo principle of unequal distribution as a basis for an already much more detailed analysis of the dynamics of capitalism, in a world where capital was more industrial (machinery, factories, etc.) than agrarian, which meant that accumulated capital could be unlimited. The main conclusion presented is often called the principle of unlimited accumulation – it is a relentless trend of capital accumulation, with an increasing share of it reaching an ever-smaller group of people, and which is not hampered by any natural barriers.

The source of many ideas of inequality is *K. Marx theory of stratification and class*. According to K. Marx, the stages of development of human history depend on the method of production. The method of production describes each economic organization, which includes technology, division of labor, and, most importantly, human relationships in the production process. These interrelationships play a central role in the Marxist class theory.

According to K. Marx, any economic organization has a ruling class that owns the means of production (factories, raw materials, etc.) and controls those means of production. In the theory of capitalism, it is the bourgeoisie (the owners of the means of production) that decides the fate of the exploited – the proletariat. K. Marx also introduced the concept of a lumpenproletariat, meaning degraded individuals, criminals, and other marginalized people who were completely excluded from the society.

According to the K. Marx's theory, the essence of the relationship between the ruling class and the exploited class is that the ruling class exploits the working class. The form of exploitation depends on the method of production. In capitalism, property owners buy the work of workers. K. Marx emphasizes that added value is created by workers. K. Marx predicted and argued that with the development of capitalism, the bourgeoisie would become even richer and the proletariat even poorer. Maturing class conflict leads to a revolution and the collapse of capitalism. However, K. Marx predictions did not come true, capitalism did not collapse.

Despite the shortcomings, T. Piketty describes K. Marx analysis as relevant to this day in several respects. Firstly, he highlighted the problem of the huge concentration of wealth in the years of the industrial revolution and tried to analyze it using all available methods. Secondly, no less important is the principle of unlimited accumulation, the essential insights of which can be applied in modern research as well. If the growth of population and output conditionally declines, accumulated wealth becomes particularly significant, especially if the situation reaches extreme and becomes socially destabilizing. In other words, low economic growth cannot offset the Marxists' principle of unlimited capital accumulation: although the imbalance will not be as extreme as K. Marx predicted, it is extremely significant. Accumulation will stop sooner or later, but the equilibrium may have reached a destabilizing level (Piketty, 2014).

S. Kuznets theory of economic development. According to the S. Kuznets theory, introduced in 1955, economic inequality in any country is defined by a bell-shaped curve ("U" curve). It has been declined that inequality increases in the early stages of industrialization, as only a minority receives the benefits and wealth generated by industrialization. In the later stages of the process, inequality decreases as more and more people can enjoy the fruits of economic growth.

According to the S. Kuznets' theory (1955), economic inequality in the world should decrease in the later stages of capitalism development, despite differences between states or economic and political decisions. And, finally, after reaching a certain level, it should stabilize. According to S. Kuznets, people have to wait patiently and the economic growth that has begun will benefit everyone.

A similar optimism is characteristic for the study performed by R. Solow (1956), who analyzes the conditions necessary for sustainable economic development. According to the study, if all variables, output, income, profit, wages, capital, asset prices, etc., progressed at the same pace, each social group would benefit equally from this growth. Thus, the latter conclusions were completely contrary to the ideas and predictions of D. Ricardo and K. Marx.

S. Kuznets' theory was the first to be based on a large number of statistics. Only in the 20th century, in the middle of 1953, when S. Kuznets' "Shares of Upper Income Groups in Income and Savings" was published, the first historical statistics on income inequality appeared. S. Kuznets described a period of 35 years (1913–1948) in only one country (the United States). However, it was an invaluable contribution based on two sources of information. The other authors of the 19th century have no chance of getting closer to US federal income tax refunds and estimates of US national income made by S. Kuznets himself several years ago. This was one of the first attempts to assess the extent of economic inequality at such a level (Piketty, 2014).

T. Malthus, D. Ricardo, K. Marx, and others have been discussing economic inequality for many decades, without quoting any sources or using any comparable methods. Thanks to S. Kuznets, objective data became available for the first time. Data collection was carefully documented in S. Kuznets' major work, published in 1953. The sources and methods used were described in such detail that each calculation could be reproduced. And most importantly, S. Kuznets confirmed his prediction that economic inequality was declining.

As history has shown, the reasons for creating S. Kuznets' curve were not correct, and its empirical basis was extremely fragile. The sharp decline in economic inequality in almost all the developed countries of the world, which was observed between 1914 and 1945, was mostly caused by the world wars. The tragic economic and political shocks, as consequences of wars, affected the richest people the most.

Since 1970, economic inequality has risen sharply in rich countries, especially the United States. Analogous trends are observed in the 21st century: income concentration in the first part of the 21st century reached the level of the second decade of the 20th century (Piketty, 2014). The rapid development of poor countries (especially China) had a major impact on reducing global economic inequality, as well as the development of rich countries during 1945–1975. However, the imbalance of the financial, real estate, and oil markets at the beginning of the 21st century raised legitimate doubts about the sustainable development of the economy. It was discussed by R. Solow and S. Kuznets, arguing that all economic variables should develop and change at the same pace.

After the Second World War, during the rapid development of economies in Western Europe and North America, economic inequality was almost forgotten. However, in the last decades of the 20th century, the issue of poverty has become relevant in the United States, Great Britain, Sweden, and other European countries. In this way, the growth strategies reducing poverty emerged, and the phenomena of growth and equality (in terms of the income distribution) were singled out as separate strategic measures influencing poverty reduction. Increasing the income of poor households became a major challenge.

The growing interest in economic inequality has been discussed in academic papers: "Economic Inequality", "On Economic Inequality", and others (Salverda, Nolan, & Smeeding, 2013). Since then, several significant studies, based on increasingly accurate data, examining the causes and concepts of inequality, measurement tools, trends, and research methods, have been conducted. It should be noted that economic inequality has been stable enough for some time.

However, in the 1980s, wage differentiation in Great Britain and the United States increased significantly and led to the emergence of new large-scale research. Studies sought to determine whether inequality exists only in the two countries mentioned or whether it is prevalent in all developing countries (Salverda, Nolan, & Smeeding, 2013). *It has become clear that income inequality between the population and households, the key aspect of all economic inequality, is growing rapidly in all countries.*

In the first decade of 21st century, it became clear that economic growth and inequality are inseparable, and economists focus on reducing extreme poverty in the past decade was not fully effective: although progress in reducing extreme poverty had been made, economic inequality continued to rise in many developing countries.

Modern theoretical approaches to economic inequality. According to the authors, *two important approaches* can be distinguished in the modern theory of inequality: *liberal – inequality is justified and it must exist; and conversely, inequality is not necessary.* Many countries have pursued economic policies based on a liberal approach (in which inequality is inevitable) until these days. Countries have to choose: either the economy grows together with growing inequality, or the economy slows down with an equitable distribution of income and wealth. And, as the economy grows, the rise in inequality will come to a halt over time and move closer to the situation in developed countries. The liberal approach says that inequality is the price we pay for economic growth. If higher economic growth is a priority, the widening of income disparities must be reconciled. However, the data from modern theoretical and practical research suggest that such an approach is questionable. The first argument is based on the theory of welfare economics and states that growing income inequality slows economic growth, and the second, the argument of proponents of the Austrian school, says that growing inequality promotes economic growth.

Theory of economic welfare. The theory of economic welfare examines how an economic activity should be managed in order to maximize economic prosperity. A. C. Pigou, V. Pareto, A. Bergson, P. A. Samuelson, K. J. Arrow, J. K. Galbraith, A. Sen, and others are known as proponents of the theory.

A. C. Pigou was one of the first to point out that economic inequality is slowing economic growth. The researcher, referring to the theory of marginal utility, suggested drawing attention to the importance of income redistribution, as transferring part of the rich income to the poor would increase the common good. V. Pareto proposed a concept of maximum benefit for society to assess changes that improve the well-being of all individuals. A. Bergson highlighted the function of social welfare – the expression of the welfare dependence on the factors determining the economy and quality of life (Daugirdas et al., 2019).

In the modern context, the EU Council (2019) defines the welfare economy as the one that:

- expands the opportunities for people to have bigger social mobility and improve life in the areas that matter to them now;
- ensures that these opportunities lead to welfare outcomes for all segments of the population, including those at the bottom of the distribution chain;
- reduces inequality; and
- ensures environmental and social sustainability.

The welfare state takes care of the well-being of its population, distributes national income fairly, and supports social satisfaction, health care, education, and other systems. Welfare state institutions take responsibility for the well-being of each individual, and a high standard of living must be guaranteed for all. This was influenced by the ideas of the prominent English economist D. M. Keynes (1883–1946), who emphasizes the necessity of state participation in a market economy. Such a provision became very popular in many Western countries, especially in Scandinavia.

The concept of the welfare state originated (matured and nurtured) in Europe seeks to alleviate or prevent social problems. However, the welfare state can exist in various forms/types. Different types of welfare states provide different social guarantees to the citizens of countries and (at the same time) generate different levels and scales of threats to social security. It has been proven that the more generous (high benefits and high quality of social services), the more comprehensive (takes into account different needs), and the all-encompassing (high coverage) activities of the welfare state, the milder negative effects of a market economy and globalization are (Aidukaitė, Bogdanova, Guogis, 2012).

The welfare state encompasses whole institutions of the state – a social policy with the social security system (consisting of social support and social insurance), education system, and institutions guaranteeing the security of citizens (police and national defense structures) play an important role. These institutions were created to reduce all forms of inequality (income, social, gender) to create equal opportunities for all citizens of the country, regardless of their gender, age, disability, racial, ethnic, religious, or other affiliation, and to ensure socio-economic security for all citizens. Also, the concept of the welfare state is inseparable from such characteristics of the 20th century as industrialization, urbanization, feminization, relatively high standard of living, solidarity between generations, and social justice. The activities of the welfare state are aimed at reducing income inequality, gender inequality, poverty, and social exclusion.

The EU Council (2019) highlighted the channels through which economic growth with prosperity supports and reinforces each other by focusing on certain policies that are essential for education and science, health, social protection, and gender equality.

It is argued that investing in people's well-being lays the foundations for stronger and more sustainable long-term economic growth. Firstly, expanding access to quality education and quality health care, as well as promoting inclusive social protection systems that promote resilience and social mobility, are shown to be powerful levers in activating the successful cycle of the welfare economy. It is also argued that, to reap the full returns, these investments must contribute to better welfare outcomes for all segments of the population. This emphasizes the importance of eliminating gender inequalities to ensure access to quality employment.

In order to ensure people have and realize greater well-being opportunities, investing in people's potential is essential. It is also a key factor ensuring long-term economic growth, societal resilience, and stability. Furthermore, only by paying sufficient attention to the sustainability of prosperity can maximize the long-term growth potential and protect the economy from adverse shocks. In both cases, the welfare economy aims to create and sustain a successful cycle in which both elements – sustainable economic growth and prosperity – work together for the benefit of the individual and society.

Austrian school. M. N. Rothbard (2019), a proponent of the Austrian school, argues that inequality is inevitable and even necessary – it works as an incentive and helps the economy to remain viable.

Representatives of the Austrian school remind us that every individual is unique, so he cannot just be "equal". In the same way, income can never be equalized. According to M. N. Rothbard (2019), equality is unattainable and is a conceptually impossible pursuit. Any effort to achieve equality is just as absurd.

Representatives of the Austrian school criticize the assumptions of the welfare economy that everyone has the right to a good life. Representatives of this school argue that everyone must take responsibility for the well-being of their lives because people cannot enjoy life without being responsible for their actions and choices. They argue that to feel self-esteem and be independent without the means to control your own life is impossible. P. Bylund (2019), criticizing the ideas of the welfare state, argues that it was they who "gave birth to a dependent society that is completely incapable of finding the values of life. It is unable to feel normal human feelings – self-esteem, nobility, and compassion. These feelings, along with the opportunities to find the meaning of life, have been taken away by the welfare state".

According to the representatives of the Austrian school, periods of the economic downturn are natural, because there is an internal mechanism in the market, a natural selection that allows the insightful ones to survive and prosper by eliminating noninsightful ones. And, the only negative consequences are systemic government intervention in the market, which paves the way for banks' credit expansion and inflation (it culminates in a phase of depression and correction).

J. E. Stiglitz: inequality is not inevitable. Columbia University professor, former President Clinton's economic adviser, head of the World Bank's economics team, Nobel Laureate J. E. Stiglitz in his report to the Roosevelt Institute on "Rewriting the Rules of the American Economy", 2015, comes up with the idea that inequality is not inevitable; it is not a social but an economic problem. It is also the reason why the economic boom (since 1990) ran slower and longer than before. Inequality is not the price for economic growth; it is rather the cause and symptom of slowing economic growth.

During the discussions at the Roosevelt Institute and the presentation of the report "Rewriting the Rules of the American Economy", which was attended by experts such as B. Solow, Nobel Laureate, economist M. Bouchley, Professor S. Johnson, Professor L. Stout, and sociologist S. Greenberg, and J. E. Stiglitz, claimed that the US Republican and Democratic administrations are guilty of pursuing policies that promote inequality. They also introduced a concept that says nothing can be done about that inequality. J. E. Stiglitz stated, "Inequality is not inevitable. It's about the decisions we make to structure our economy".

One of the key questions that J. E. Stiglitz raises is why inequality is such an important issue?. The answer, according to J. E. Stiglitz, is simple – because it hampers economic growth, and at the same time it harms the well-being of people, but the causes of inequality are much more complex.

The main conclusions of the J. E. Stiglitz report are, firstly, inequality is a problem of political economy and it is closely linked to political decisions. Secondly, the essence of the problem is not some individual solution, corporate tax, health reform, or labor standards, but their totality. "Our economy is a system", says J. E. Stiglitz. "Tackling inequality requires a systemic approach in many areas: finance, corporate governance, executive pay, tax policy, antitrust, monetary policy, education, health, and labor law".

"Inequality is not an inevitable fate, it is the result of political decisions" (Stiglitz, 2015). This is evidenced by the experience of some countries that have succeeded in reconciling equality with economic growth. These are the Scandinavian countries, Singapore, and Mauritania, which have primarily invested in education and science.

Options proposed by J. E. Stiglitz that should reduce inequality are

* investment in research, infrastructure, and education, increasing access to higher education;
* the introduction of a minimum wage (US profits have grown disproportionately to wages in recent years). Such imbalances in income distribution are a source of inequality; and
* tax policy needs to be more progressive and balanced.

Inequality is not the result of economic growth but, on the contrary, the cause of slowing economic growth. The higher the inequality, the lower the economic growth (Stiglitz, 2015). It should also be noted that inequality has a negative impact on the well-being of human life. Research shows that the distribution of the created product (while ignoring social justice) has a negative impact on the quality of life of the population – demographic processes, health, a material situation of people, and reduces access to education, and training. In this context, extreme inequality is treated as excessive inequality (not any other inequality that promotes development and competition itself).

According to J. E. Stiglitz (2015), the greater wealth exclusion among members of society is the less rich want to invest in building community well-being. The rich part of society simply does not need the state to invest in education, medical care, or personal security – all of this they can buy themselves. Over time, the exclusion of middle class and rich people increases, groups lose any empathy they had for another group. Rich people can only be concerned about a strong government that can balance society, "take away" part of their wealth, and donate it to the common good of society.

According to J. E. Stiglitz (2015), nearly a quarter of the nation's total income goes to the 1% of the richest Americans each year. It is also important to mention that the incomes of 1% of the richest US population have increased by 18% over the past decade, while the incomes of the middle class have declined. In addition, men with a secondary education felt this particularly sharply – their income fell by 12% in the last quarter of the century. Thus, income growth in recent decades has been felt only by the richest people in the country.

Economists still look for ways to explain growing inequality in the United States and around the world. Firstly, the development of technologies focused on saving human resources has reduced the need for middle-class workers. Secondly, globalization has influenced the emergence of a global market and boosted the opportunities for workers to migrate and work in other countries. Thirdly, social change has also been affected, e.g., by the decline of trade unions, to which a third of all Americans have recently belonged (only 12% belong now).

However, according to J. E. Stiglitz (2015), *inequality exists mainly because 1% of the country's population seeks for it.* Proof of this may be tax policy. Increasing tax incentives for taxable capital gains, which allow for the accumulation of wealth, have created particularly favorable conditions for certain groups in society. Monopolies and partial monopolies have always been a source of economic power – from John D. Rockefeller at the beginning of the last century to Bill Gates at the end of the last century. The scale of today's inequality is mostly the result of antitrust policies' absence and manipulation of the financial system, which has been ensured by changes in the rules made by the financial industry itself. The government lent money to financial institutions at near-zero interest rates and provided financial assistance on favorable terms when those institutions began to collapse.

J. E. Stiglitz (2015) emphasizes that economic globalization benefits only the rich: it encourages countries to compete for business influence, reduces health and environmental protection, and restricts the so-called fundamental labor rights (which include the right to bargain collectively). We can only try to imagine what the world would look like if economic globalization encouraged countries to compete and compete for workers. The authorities of the countries would compete on which of them will ensure better economic security, reduce taxes for workers, provide better education, and guarantee a cleaner environment – all the things that most of the people care about.

The difference between 1% of the richest and 99% of the rest of the population's annual income is huge, and in terms of accumulated capital and other values it is even bigger. However, the greatest of all the losses that society suffers from inequality is the loss of awareness of fairness and behavior, decency, equal opportunities, and a sense of community.

J. E. Stiglitz argues that there is no major controversy over the fact of growing inequality. However, it can still be argued that inequality is a fundamentally positive thing: if the income of the rich group gradually increases, so does the income of all the rest of the population. This argument, according to J. E. Stiglitz is wrong: as long as a small proportion of the rich group accumulate wealth, the majority of the population cannot ensure normal living conditions. An ordinary US worker receives the same pay for a day's work as he or she did 3–4 decades ago. In this way, inequality turns into economic dysfunction (Stiglitz, 2015) – hampers economic growth. According to the economist, it is no coincidence that the period when most Americans received higher net incomes – when inequality was reduced in part by progressive taxes – was also a period when the US economy was growing rapidly. It is not a coincidence that the current recession, such as the Great Depression, has provided opportunities to increase inequality.

The movement of money from the "bottom" to the "top" of society reduces consumption, as individuals with higher income spend a relatively smaller share of their income than individuals with lower income. At first glance, this fact may seem wrong, because the rich spend money as much as they do. However, even if Mitt Romney, whose revenue has reached 21.7 million dollars in 2010, decides to afford more than usual, he would spend only a small portion of his income a year in order to support his family and a few homes he owned. However, dividing the same amount of money by 500 people – e.g., giving each of them a job and paying $43,000 a year – will make it clear that almost all the money will be spent. There is a rule – *the more pronounced the differentiation is, the more the common demand decreases.*

If someone does not change and stop it, common economic demand will be lower than supply, which means that unemployment will rise, reducing demand even more. In the 1990s, that "someone" was the bubble of Internet companies, at the beginning of the 21st century – the bubble of real estate.

The concept of justice is subjective. People are not angry or jealous of the accumulated wealth of people who have changed the economy for the better – computer inventors and pioneers of biotechnology. But most of these people are not at the top of our economic pyramid. In a society where inequality is on the rise, justice is not just measured in terms of wages, income, or wealth.

According to J. E. Stiglitz (2015), justice is a much more complex concept. *One of the most important conditions for justice is an opportunity: everyone must be able to realize their higher needs.* Lack of opportunities has many consequences. Growing inequality simply undermines confidence. It is much easier to act together when all individuals are in a similar position.

J. E. Stiglitz emphasizes that there is no good explanation why 1% of the richest (with good education, excellent counselors, and insight) are so ill informed. The richest of past generations knew they would not be able to feel safe on their own if society was unstable. For example, H. Ford realized that the best he could do for himself and his company was to pay employees a decent salary. He wanted his employees not only to work hard but also to be able to afford to buy the cars he produced.

F. D. Roosevelt realized that in order to preserve capitalist America, it was necessary not only to distribute wealth through tax and social security systems but also to curb capitalism itself. Proponents of capitalism disliked neither F. Roosevelt nor economist J. M. Keynes, but they managed to protect capitalism from the capitalists. R. Nixon realized that the investment would ensure social well-being and economic stability; therefore, he invested constantly: in "Medicare", "Head Start", the social security system, and efforts to put it in order.

J. E. Stiglitz argues that *inequality is the result of economic policies.* The concentration of tangible assets "on the one hand" not only destroys the principles of social justice but also limits the opportunities of the remaining members of society and their quality of life. It also inhibits and does not promote the socio-economic progress of the state itself. The roots of the problem lie not in a single fragmentary factor but depend on a combination of factors. Tackling inequality requires a systemic approach, an integrated solution in many areas, from financial and antitrust reforms to education and employment regulation.

R. Reich and his basic economic assumption. In his book *Aftershock: The Next Economy and America's Future*, 2010, the author convincingly shows that the cause of the global economic crisis is not public debt, but the growing socio-economic inequality that transcends limits of economic security. The neoliberal, monetary model of economic globalization results in the uneven development of the world economy, which favors economically strong countries and large corporations, and negatively affects the positions of less developed countries and small as well as medium-sized enterprises. However, global research shows that socio-economic inequality is a profound problem that is becoming a brake on global economic development, not only in the weaker developing countries but also in the leading countries, making socio-economic inequality the cause of the economic crisis.

Recent research has shown that, in fact, US household incomes have been averagely growing at all-time only due to the growth of more affluent household incomes. And, for the rest – only debt grew exclusively. In reality, the average income of the US population has been declining since the 1970s. An objective analysis of the income of the population allows us to draw the main conclusion: the credit boom of the last 30 years has been accompanied by a widening gap between the income of the richest people in the United States and the rest of the population.

R. Reich argues that the big crisis of 2008 is the result of a growing misallocation of income. Instead of new strategies to enable the middle class to successfully deal with these difficult times, political leaders (who blindly believed in a free market) opted for privatization and reduced state influence on the economy, cutting taxes on the rich, and destroying the social security system. The result is a frozen wage for most Americans, declining employment guarantees, and growing inequality. The results of economic growth are also being used by fewer and fewer people.

At the end of the 1970s, 1% of the richest Americans disposed of 9% of total revenue. In the 2007s, 1% of the richest already disposed of 24% of national income. If most people's incomes are growing poorly, the middle class can only meet its quest for a better life by getting deeper into debt. Consumption stimulates the economy and creates jobs, but, in this case, the process does not take long. It is impossible to keep borrowing forever. At some point, it is shown in the 1929s and 2008s – the time "to pay for it" comes. *When the equitable distribution of income collapses, the economy needs to be rebuilt for the middle class with sufficient purchasing power.*

R. Reich argues that *the concentration of income in small sections of society has a negative impact on the economy*. Until the 2008s crisis, half of all consumers' spending belonged to 20% of citizens with the highest incomes, 40% of these expenditures went to the upper decile of the population. However, all this was not related to spending, it is related to the fact that 10% of the population disposed of 50% of total income.

According to the R. Reich, if the dominant middle class received a larger share of income, the volume of consumption would be much higher and this class would not be so indebted. If most Americans received a proportionate share of the country's total income, they could live not worse, and even better than before: to put off saving for the "black day", successfully find a way out after losing their jobs or having their wages fall. In that case, they would not have to borrow as much.

The problem did not lie in the fact that Americans did not live within their means – their income simply did not follow the perfectly correct logic of what a country can afford in a thriving economy.

The first stage of the American capitalist economy (1870–1929) was characterized by an increase in the concentration of income and capital, and in the second stage (1947–1975), wealth was distributed more or less evenly. In the third stage (1980–2010), there was again a large concentration of capital. R. Reich says, "we need to start the fourth stage, in which distributed wealth becomes the norm again".

In the 1945s, the rapid economic growth began and lasted for 25 years. Much of the tax revenue came from the middle class and the richest, with a tax burden ranging from 79% to 94% during the years of war. In the 1950s, this figure was 91%; in 1964s, 67–77% (maximum tax rate). Thus, the high income had to give far more than 50% of the income (after all benefits and credits). High tax levels did not mean a contraction of the economy at all; on the contrary, the middle class prospered and fed the economy.

The basis of America's great prosperity is an economic policy, which was radically different from the one that led to the crisis in the 1930s. Through prosperity, the government used Keynesian methods to achieve full employment, provide social security, and increase public investment. As a result, the share of gross income for the middle class increased. But, here's what's interesting – the economy has grown rapidly, and everyone has benefited from it. Equal distribution of income was fully compatible with economic growth and became a condition for economic growth.

Since the 1970s, the US government has been engaged in deregulation and privatization for three decades in a row. It has increased the cost of higher education, reduced the number of education and pre-employment programs in colleges and universities, and reduced the provision of public transport and other public services. The government has almost abolished the social security system – the funding of programs for unemployed families with children has been reduced, and the procedure for providing unemployment insurance has been tightened; therefore, in the 2007s, the funding was received by only 40% of the unemployed people. It has halved the tax rate on wealthy sections of the population, allowed many wealthy Americans to formalize their income as a return on invested capital and pay only 15% of the taxes. It has also reduced the inheritance tax, which affected 1.5% of the rich people population. And, at the same time, it increased sales taxes and personal income tax, which accounted for a significant share of middle-class income. Companies were allowed to fire workers and reduce wages easier; it also allowed them to reduce benefits and transfer risk onto the shoulders of workers.

The problem was not the fact that Americans did not live within their means, but that their incomes did not increase. The main task of the economy is to increase the wages of the middle class, so it receives sufficient income to be able to fully consume and purchase the necessary goods and services. Nevertheless, it is impossible to overcome unemployment, increase budget revenues, and stimulate economic growth. Differentiation of income and wealth threatens the unity and cohesion of society, destroys democracy, and destabilizes the economy. Instead of reducing wealth inequalities, the US monetarists began rescuing banks and the financial system. However, until the focus is on an equitable distribution of income and wealth, the needs of the middle class are met, the economy does not recover, and the threat of a crisis remains.

T. Piketty and two concepts of inequality. In the book *Capital in the Twenty-First Century* (Piketty, 2014), T. Piketty concludes that wealth and income inequality are not only the result of economics but also of politics.

> Economic determinism should be avoided when it comes to wealth and income inequality. The roots of the history of the distribution of wealth have always been in politics and cannot be attributed solely to the operation of economic mechanisms. Especially bearing in mind that the decline in inequality (that took place in the most developed countries between 1910 and 1950) was primarily a consequence of war and policies. The resurgence of inequality (after the 1980s) has similarly been caused by political changes over the last few decades, especially in the areas of taxation and finance.

According to T. Piketty, the history of inequality is shaped by the assessments of actors in economic, social, and political life – they decide what is right and what is not. The relative power and ultimate collective choice of those actors also play a big role in this shaping. Therefore, the history of inequality is a common product of all actors' activities.

The second conclusion is that the dynamics of asset distribution reveal powerful mechanisms that reflect the alternating processes of convergence and divergence. According to T. Piketty, there is no natural, spontaneous process that would prevent the permanent establishment of destabilizing, inequality-promoting forces.

T. Piketty puts forward two important and interesting hypotheses about *"growing human capital"* and *"the struggle between the generations to fight class changes"*.

Firstly, the main drivers of convergence in reducing inequalities are knowledge dissemination, investment in education, and capacity building. The law of supply and demand, as well as a variant of this economic law – the movement of capital and labor, always pushes for greater convergence. However, the impact of this economic law is less than the dissemination of knowledge and capacity building; therefore, the contradiction in its consequences is often unclear.

Dissemination of knowledge and capacity building is the key to productivity growth and the reduction of inequalities both – within and between countries. This is evidenced by the progress of former developing countries (led by China). Currently, developing countries are catching up with the most developed ones. By adopting the production methods and capacity building of developed countries, developing countries have increased their productivity and national incomes. According to T. Piketty, the process of technological convergence has been driven by borders open to trade, but it is essentially a process of disseminating and exchanging knowledge – the takeover of public goods, not the effect of a direct market mechanism.

From the point of view of theoretical rationality, there may be other forces that promote greater equality. It could be assumed, e.g., that over time, production technologies require more sophisticated skills from employees, the ability to increase the share of labor income as the share of capital decreases: this phenomenon could be called the *"growing human capital hypothesis"*. In other words, technological progress should automatically lead to the victory of human capital over financial capital and real estate, the victory of talented executives over rich shareholders, and the victory of skills against nepotism. Thus, inequality would become more meritocratic and less static (not necessarily less): economic rationality should, in a sense, automatically lead to a democratic rationality.

The second hypothesis of T. Piketty is that the *"class struggle" will automatically give way* – due to the increase in life expectancy in recent years – to the *"generation struggle"* (which is less divisive because we are all young, at first, and then – old). In other words, this inevitable biological fact should mean that the accumulation and division of wealth no longer anticipate the inevitable conflict between tenant dynasties and dynasties that cannot offer anything but labor. The logic is to save for life: people accumulate wealth when they are young to take care of themselves at old age. It is argued that advances in medicine, along with improving living conditions, have fundamentally transformed what we call capital.

T. Piketty argues that, unfortunately, these two optimistic beliefs (the human capital hypothesis and the class struggle change of generations) are fundamentally deceptive. This type of transformation is both logically possible and to some degree real, but its impact does not have as profound consequences as one might imagine. There is little evidence that the share of labor income in national income has increased significantly over a truly long period of time: in the 21st century, "nonhuman" capital seems to be as irreplaceable as it was in the 18th or 19th century, and there is no reason why its importance should not grow even more. Now, as in the past, wealth inequality is first and foremost seen in age cohorts, and inherited wealth at the beginning of the 21st century has the same decisive power that Balzac's "Father Gory" had. However, *in the long run, the driving force behind greater equality has been the dissemination of knowledge and capacity building.*

T. Piketty emphasizes,

> The truth is that economics should never have been separated from other social sciences. It can only develop in conjunction with other sciences. If we want to learn more about the historical dynamics of the distribution of wealth and the structure of social classes, it is clear that we need to think pragmatically and use the methods of historians, sociologists, and political analysts, and economists. We need to start with the essential questions and make an effort to answer them. Disputes about the purity of the subject and struggles for influence are important a little or have no importance at all.

Economic inequality's impact on socio-economic progress. Economic inequality's impact on socio-economic progress is not unambiguously defined. It is confirmed by the abundance of empirical research, the indicators used in research, and the contradictory evidence. Constructing different econometric models while analyzing the impact of economic inequality on this particular topic, the authors often use similar statistical methods (regression analysis using panel data). However, they incorporate different variables in the model often and interpret the results ambiguously.

Based on this point of view, research projects could be divided into several areas:

- Representatives of the first direction research projects highlight *the negative relationship between economic inequality and socio-economic progress* (Alesina& Rodrik, 1994; Alesina & Perotti, 1994; Atkinson, Piketty, & Saez, 2011; Persson & Tabellini, 1994; Bourguignon, 2004; Rakauskienė, 2015, 2017; etc.).
- Proponents of the second direction argue that economic inequality *has a positive impact on socio-economic progress and GDP growth* (Rothbard, 2019; Bylund, 2019; Li & Zou, 1998; et al.).
- The third line of research argues that *there is a neutral relationship between economic inequality and socio-economic progress* that impacts fluctuate depending on circumstances and factors (Kuznets, 1955; Barro, 1999, 2000; Halter, Oechslin, & Zweimüller, 2013; Dominicis, Florax, & Groot, 2008; Hoeller, Jourmand, Pisu, & Bloch, 2012; Dabla-Noris et al. 2015; Grifell-Tatje, Lovell, & Turon, 2018; et al.).

Only a few of the most important authors will be mentioned in this book.

A. C. Pigou (1920) was one of the first to point out that economic inequality slows down economic growth. The researcher also emphasized that national income cannot accurately reflect the well-being of the society, because, in real welfare, externalities usually have no monetary expression. Therefore, the real level of well-being may arise without a rise of quantifiable well-being.

S. Kuznets curve. The beginning of more detailed research on the interaction between economic inequality and economic growth is linked to the 1955's study paper by S. Kuznets mentioned above. The main question the researcher sought to answer is whether the economic growth of states has a positive or negative effect on economic inequality. While describing the nature of the relationship between economic growth and economic inequality, S. Kuznets (1955) asserts that, at the stage of a country's economic development, economic inequality increases along with rising incomes. Nevertheless, as incomes continue to grow, economic inequality begins to decrease. In this way, an inverted letter "U" curve was presented to illustrate the hypothesis put forward by S. Kuznets.

As history has shown, economic inequality and its extent changed over time: in the early stages of economic growth, when the agrarian social order was replaced by industrial, economic inequality increased. After that, it became stable until the late stages of economic development, when it started to decrease again. The latter fluctuations were most common in the older states, where the growth of the modern economy had the greatest impact. However, the same tendencies were observed in other states as well.

S. Kuznets emphasized that economic inequality and economic growth are linked closely. Long-term fluctuations in economic inequality are also closely linked to the extent of capital formation – higher inequality corresponds to a higher level of national savings while lower inequality – to a lower level.

S. Kuznets examined how the experience of developed countries can help the economy of less developed countries with a lower average per capita income. In such countries, the first decile cannot be less than 6–7% of average income; otherwise, the latter group of society would simply not survive. Meanwhile, in developed countries, where the share of average per capita income is higher, it is enough for the first decile to be 3–5% of average. In this case, it would not condemn the latter group of society. Almost all the counties feel constant pressure to improve the situation of the lowest income groups, it is clear that the overall level of the country's total income and income distribution needs to be taken into account.

Currently, the International Monetary Fund (IMF) experts (2015) also emphasize that economic policymakers need to pay more attention to the poor and the middle class for two reasons. Firstly, a more even distribution of income – increasing the share of income allocated to the poor and maintaining the income level of the middle class – has a positive effect on economic growth. Secondly, inequality develops in different directions when it comes to developed and lagging countries.

S. Kuznets noted that in less developed countries, the middle class does not seem to exist: there is a huge contrast between the population with an income lower than the average of the country and the population with a high income. The first group, most of the time, is significantly larger. According to S. Kuznets, developed countries are characterized by a more gradual rise from the lowest to the highest. In this case, the majority of the population earns more than the average income and the richest group has a lower share of income compared to the richest in less developed countries. The situation totally differs in less developed countries, which are characterized by significantly higher income inequalities. The richest members of these societies have a relatively high share of income. Meanwhile, other parts of the population (in the lower than the tenth deciles) receive a share that is lower than the low average in the country.

The greater inequality in the distribution of total income that exists in less developed countries is associated with a much lower level of per capita income (Kuznets, 1955). Such an inequality has consequences. In particular, its impact is much stronger and particularly pronounced in less developed countries. Living in material poverty and trying to achieve an already low average income level in the country causes psychological barriers and prevents human self-realization, which is also significantly more pronounced in less developed countries. In addition, people in less developed countries have the opportunity to save only if they earn above-average incomes. According to S. Kuznets, in developed countries savings can already be accumulated by the representatives of the fourth quantile, but in less developed countries savings can be accumulated only by representatives of the top of the "pyramid". In this case, it becomes clear that income and wealth concentration disparities are greater in less developed countries. Similar trends persist to this day.

R. Barro (1999; 2000) examines the inequality factor based on S. Kuznets main ideas and the "U" curve. Economic development is based on changes in society's order when the agrarian society became industry-oriented. The incomes of people experiencing this transformation increased which also contributed to growing economic inequality – "the rich" group, especially representing the industrial and urban development sectors, grew. Thus, in the early stages of the latter process, the relationship between per capita resources and the extent of inequality was positive. As the agricultural sector gradually declined, the poor and those working in the agricultural sector were given more and more opportunities to join the industrial sector. In addition, the prevailing trend was that most workers who started at the lowest level of the industrial sector moved up the career ladder. In turn, since the agricultural workforce got weaker, it led to a relative increase in income in this sector. The latter factors, interacting together, reduce the overall inequality rate. Therefore, in the later stages of economic development, the relationship between per capita resources and inequality became negative. Thus, R. Barro (2000) confirms the basic idea of S. Kuznets (1955) that as the economy develops, inequality first increases and then decreases.

On the other hand, R. Barro (1999) points out that the less advanced sector may use older technologies, while the more advanced use new and innovative methods. The transition from old to new technologies requires a certain amount of preparation, acquaintance, and reorientation. From this point of view, most of the technological innovations such as factory development, electricity, computers, the Internet, etc. – were often increasing inequality. This was also largely influenced by the fact that a certain small group of people shared a large income, which was generated by a technologically advanced sector. As people move to a more advanced sector in the process of transformation, inequality increases along with the rising rate of resources per capita. However, the more people take advantage of advanced technology, the faster inequality decreases. This phenomenon exists because relatively few people are left out (in a less advanced sector), but the wages of the latter may also increase as the supply of resources in this sector decreases (Barro, 2000).

Empirical research by R. Barro (2000) suggests that economic inequality is not unambiguously linked to economic growth and the scale of investments. It is noticeable that inequality hampers economic growth in poor countries, but in rich countries, on the contrary, inequality is exactly what stimulates the economy. In addition, the level of economic inequality must be taken into account when assessing the impact of economic inequality on economic growth.

R. Barro (2000) emphasizes that economic inequality affects economic growth in several ways. First, the impact of the *shortcomings of credit markets* is highlighted. Limited borrowing opportunities mean that often the use of investment opportunities depends on an individual's income and assets. For example, growing economic inequality reduces the incomes of the poor class, making higher education less accessible for the latter, unless they are allowed to invest in human capital – to borrow (credit would be granted). However, such a possibility is often limited.

Secondly, there's an exceptional impact *through fiscal policy*. It is argued that higher levels of inequality encourage redistribution (changing the balance of available resources by redistributing them from the rich to the poor) through political processes – through social benefits, taxation of labor income, etc. With declining incomes, the rich may invest less, so economic growth slows down with reduced investment. Higher taxation of the rich may also contribute to the slowdown of economic growth. Thus, as higher economic inequality leads to greater redistribution, economic growth declines accordingly. R. Barro (1999) points out that the rich can prevent various actions of redistributive policy by lobbying or even buying the votes of members of the legislature. However, lobbying activities would consume more resources, encourage government corruption, and, consequently, hamper economic development. Thus, even if no property redistribution actions are taken in the state, inequality can negatively affect economic growth through certain political processes. On the other hand, government spending on the poor can also boost the growth of the economy. Therefore, the impact of economic inequality on economic growth through fiscal policy is ambiguous.

Thirdly, the *impact of socio-political unrest* is highlighted. R. Barro (2000) argues that economic and such property inequalities force the poor class to commit crimes, stirring up, or interact in other activities that disrupt the common good. As a result, various property rights threats hinder investment and, consequently, economic growth. Finally, no less important is *an aspect of savings.* Rising economic inequality increases the savings potential of the rich, and the accumulated funds can stimulate investment in technological development, which offers economic growth.

In summary, the economic impact of inequality on economic growth cannot be unambiguously defined. Economic inequality can have a positive effect on economic growth through savings, but credit market failures, fiscal policy, and socio-political unrest can also have a negative influence on growth.

A. Atkinson, T. Piketty, and E. Saez (2011) analyze income distribution based on income tax statistics in more than twenty countries. Research has shown that a small percentage of the population receives the highest income, but they make up a very large proportion of total income and taxes paid. Thus, overall economic growth and indicators of economic inequality depend on the income share of the rich group.

The model of A. Atkinson, T. Piketty, and E. Saez highlights these empirical results. In the first half of 20th century, a lot of countries experienced significant declines in revenue. The largest drop in wealth in these countries was often explained by World Wars or the Great Depression. However, in some of the countries which were not directly involved in World War II, the decline in income was more moderate. In all the countries with income structure data, the income of the upper deciles was mainly capital income (as opposed to labor income). Therefore, the decline in the highest deciles was due to a decrease in the concentration of wealth of the rich population. In contrast, in the first half of the 20th century, the falls in the higher-income groups (which are below the highest decile such as 4%, and mainly made up of labor income) were much smaller than those of the richest.

Until 1949, the income distribution of the highest decile in all studied countries was small. In the second part of the century, the income of the highest decile changed according to the "U" curve – income declined for five decades and then began to grow. The share of rich population income in all Western English-speaking countries (Europe, North America, Australia, and New Zealand), as well as China and India, increased significantly. During the same period, southern and northern European countries also experienced income growth in the highest decile, albeit at a lower rate than in English-speaking countries. Meanwhile, among continental European countries such as France, Germany, the Netherlands, and Switzerland, just a very small increase in rich incomes was recorded.

Research by A. Atkinson, T. Piketty, and E. Saez suggests that the rich can have a significant impact on overall economic inequality. Studies have shown that 1% of the US richest population increased the Gini coefficient by 8.4 percentage points from 1976 until 2006. This is higher than the officially announced increase in Gini from 39.8% up to 47.0% during the period of 1976–2006. Although the analysis assessed the role of the rich population only at the national level, it is clear the result is important globally.

H. Li and H. Zou (1998) argue that economic inequality can theoretically lead to higher economic growth because redistributive policies encourage investment in the public sector, which benefits the majority of society. In this regard, researchers argue that economic inequality is positive and most of the time highly related to economic growth. Research shows that the more evenly distributed incomes are, the higher income tax rates are, which leads to slower economic growth. For example, for an egalitarian society (proponents of universal equality) living in a poor economy could be very difficult to influence economic growth. Conversely, an economy characterized by a very unequal distribution of income could influence economic growth through an increase in the wealth of the rich.

D. Halter, M. Oechslin, and J. Zweimüller (2013) argued that higher inequality contributes to short-term economic performance but reduces GDP per capita growth in the future; therefore, the long-term effect is negative. According to research by Barro (2002), Alesina and Rodrik (1994), Alesina and Perotti (1994), and Persson and Tabellini (1994), economic inequality can influence economic growth through different areas (channels). Therefore, it can be argued that the impact of economic inequality on growth is complicated and complex: higher inequality can have positive and negative consequences for the economy (Halter, Oechslin, & Zweimüller, 2013). In addition, some of these effects usually occur rapidly (short-term effects), while others may affect growth only in the long term (subsequent [delayed] effect).

Research performed by Halter, Oechslin, and Zweimüller (2013) highlights that higher inequality helps to grow in the short term, but can be detrimental in the longer term. In this context, researchers explain that positive and negative effects on economic inequality depend on the way (channel) it is affected. The growth-enhancing effects stem from purely economic mechanisms (savings opportunities, imperfections in capital markets, promotion of innovation) and are therefore identified relatively quickly. Meanwhile, the dampening effects on economic growth are related to the political process, and the roles of social and political movements. It could also be influenced by changes in the education of society and these effects only occur much later. The results of this empirical research have confirmed that the growth of economic inequality usually has a positive effect on the average GDP growth rate per person over a five-year period, but eventually, the effect of economic inequality becomes negative when the ten-year period is analyzed.

The results of Malinen's (2013) research show that the impact of economic inequality on economic growth is statistically significant and negative when a new income distribution indicator – the Estimated Household Income Inequality (EHII) dataset 2008 – is used. The researcher criticizes the idea to use Gini's index for the availability of the data and chooses the EHII 2008 measure of economic inequality for his research. The use of the EHII 2008 has increased the coverage of data in general: previous studies have typically been able to use data from 40 to 50 countries, while the latter has been measured using information from 70 countries (Malinen, 2013).

The measure of EHII 2008 inequality is only the presentation of statistical summaries (analogous to the Gini coefficient). Thus, the level of economic inequality, measured by the EHII 2008, may not reflect the true level of inequality in some countries. Furthermore, the results describe only the short- or medium-term relation between economic inequality and economic growth, so the long-term relation remains unexplored. However, according to Malinen (2013), despite the shortcomings of the EHII 2008 results, the findings of the study can be considered more reliable than many previous studies. This is due to inaccuracies and inconsistencies in the Gini coefficient.

Dominicis, Florax, and Groot (2008) state that *economic theory does not unambiguously predict the impact of economic inequality on the direction of economic growth.* However, the direction and magnitude of the relationship between economic inequality and economic growth are important for policy decisions and policy evaluation. Dominicis, Florax, and Groot (2008) conducted a meta-analysis of more than 400 estimates of the economic inequality impact on economic growth. They found that different forecasting methods, data quality, and sample size have ambiguous effects when measuring economic inequality's impact on economic growth. For example, research that relies on a fixed-exposure model systematically provides higher estimates of exposure magnitude. The use of fixed-impact methods, as well as the introduction of regional-specific variables, reduces the negative impact of inequality on economic growth outcomes and emphasizes the positive impact. It is also observed that the unequal distribution of income has a more negative impact on the economies of less developed countries. In addition, the length of the period of economic growth in question has a significant impact on the results: the longer the period of economic growth is in question, the fewer coefficients that can measure it exist. This result confirms that the relation between economic inequality and economic growth differs in the short and long term. Researchers also note that the quality of data on income distribution is a key factor. When authors use data which quality is relatively poor, the impact of economic inequality on economic growth (positive or negative) is weaker. In addition, the inclusion of additional inequality indicators (such as land inequality or inequality in human capital) has a significant impact on the size of the estimates. It can therefore be concluded that the use of consistent inequality data is necessary to measure the impact of economic inequality on economic growth.

Bourguignon (2004), like Alesina, and Rodrik (1994) and Atkinson, Piketty, and Saez (2011), emphasizes the importance of redistributive policies in promoting economic growth through the gradual reduction of economic inequality and poverty.

Bourguignon (2004) argues that redistributing wealth from the rich to the less affluent can have a positive impact on economic growth. It should be emphasized that a redistribution of wealth rather than redistribution of income can lead to the desired positive effects on economic efficiency and growth. Meanwhile, income redistribution can have the exact opposite effect on growth. By reducing the expected return on the acquisition of physical and human capital, it can distort the economy and reduce savings and investment, and hence growth. In order to have effective economic growth and stimulate it even more, redistribution should be related to wealth, but not to income or consumption expenditure.

On the other hand, it is questionable whether the reallocation of assets is possible in reality. The redistribution of wealth can only take place in exceptional circumstances, which are often linked to "political violence" (e.g., land reforms, expropriation of property, subsidized transactions, etc.). Thus, the feasibility of a redistribution of wealth largely depends on the political context. However, such a redistribution of wealth will be opposed by the richest members of society, making such a redistribution an unrealistic option.

Bourguignon (2004) emphasizes that, in a general sense, redistribution is necessary for growth. For example, less affluent people can become educated when the incomes from richer and often more politically active members of society are redistributed and invested in education, etc. – activities relevant to the public good. However, the initial level of per capita income (initial economic inequality), which has a positive or negative effect on economic growth, is also particularly important.

Hoeller, Journard, Pisu, and Bloch (2012) emphasize that, from a theoretical point of view, the impact of economic inequality on economic growth is ambiguous. First, inequality can have a positive effect on growth: (1) through the concentration of rich savings; since investments depends on the size of savings, higher inequality can lead to faster growth; (2) the concentration of assets encourages the creation of new activities; and (3) incentives to work more productively are stronger in unequal societies. However, secondly, the mechanisms underlying the negative relationship between inequality and growth need to be highlighted: (1) a domestic fiscal policy: more unequal countries redistribute more, leading to market distortions and lower growth; (2) social and political instability: high economic inequality contributes to political and social instability as more people engage in illegal activities, such as crime or violent protests, which hamper investment; and (3) credit market imperfections: such shortcomings lead to underinvestment in human capital.

Dabla-Noris et al. (2015), based on the IMF, reveal the importance of economic inequality for economic growth and emphasize that distribution of income is important for growth. Researchers argue that increasing the income share of the poor class and maintaining incomes of the middle-one has a positive impact on growth through a variety of relevant economic, social, and political channels. It should also be noted that economic inequality operates in various directions and it largely depends on the country's policies, institutional environment, and other driving forces. Researchers note that *the focus of politicians and the desire to increase the incomes of the poor class and middle-class members of society could boost economic growth.*

Dabla-Noris et al. (2015) emphasize the established inverse relationship between the share of income of the rich (20% of the rich group) and economic growth. If the income share (of 20% people who belong to the richest group) increases by 1% point, GDP grows 0.08% points less over the next five years. It indicates that the benefits of such an increase are not being felt in the lower stages. However, if we increase the income of 20% of the poorest people, a similar increase could lead to 0.38 percentage points higher economic growth. This positive relationship between disposable income and higher growth remains in the second and third quantiles (middle class). This result remains even after the reliability tests and corresponds to the latest results obtained in a smaller sample of developed countries (Dabla-Noris et al., 2015).

In 2015, *A. Deaton* was awarded the Nobel Prize in Economics for his contribution to the analysis of consumption, poverty, and well-being. A. Deaton's methodological approach is based on data concerning individual experience and household behavior. According to the researcher, this decision allows to achieve more adequate results, which are much more important than standardized macroeconomic models. According to Deaton (2003), in order to develop effective economic policies that promote prosperity and reduce poverty, a good understanding of consumption principles – at the individual level, at the level of rich and poor households – is a must. According to the researcher, understanding the levers of individual consumer choice helps to create economic policies that promote the prosperity of society and the reduction of poverty. In the process of globalization, recurrent economic crises highlight the shortcomings of a system that has failed to deliver economic growth and quality of life.

Grifell-Tatje, Lovell, and Turon (2018) singles out the components of socio-economic progress: (1) growth of productivity and the value it adds, (2) a balanced distribution of created value, (3) the reuse of resources that have been used for productivity growth, and (4) the creation of a positive social effect. The aspect of equality (inequality) is closely related to all the identified components of socio-economic progress.

Grifell-Tatje, Lovell, and Turon state that even a balanced distribution of created value does not guarantee that business contributes to the reduction of economic inequality. This relates to Davis's (1947) "functional" distribution of income among those stakeholders who directly contribute to value creation – performs a certain function. However, in order to reduce economic inequality, a balanced distribution of value must be addressed to society as a whole as a stakeholder.

The impact of the identified components of socio-economic progress on economic inequality cannot be unambiguously defined. Several examples support this, such as:

- Two groups of consumers can be distinguished: one who buys essential goods and the other who buys luxury goods. Even if the prices of luxury goods fall by exactly the same amount as the prices of basic goods, economic inequality would increase – the wealthiest members of society who buy luxury goods generally improve their living standards and can buy even more luxury goods compared to consumers of basic goods.
- Two types of payment for work performed can be distinguished: one who receives a fixed monthly salary and other who are paid an hourly rate for the work. It is not uncommon for employees with a fixed salary to earn more than those who are paid for exact hours. Therefore, if the incomes of both excluded groups increased equally, economic inequality would increase. Similar trends emerge with regard to the gender pay gap. The Organisation for Economic Co-operation and Development (OECD) (2004) emphasizes that economic inequality has risen sharply since the 1980s, precisely because of the widening pay gap between high-paid and middle-paid workers.
- The impact of technological progress on economic inequality is also debated widely. It is often assumed that technological progress requires new technological skills from workers and thus contributes to wage inequality. However, research suggests that new technological skills are not the only explanation for the wage gap and growing economic inequality. In addition, technological innovation has been expected to lead to the loss of many jobs (especially due to automation) and thus increase economic inequality. However, researchers emphasize that technological advances allow only certain tasks to be taken over and automated, but never replace the worker.
- With regard to capital investment, the OECD (2014) notes that incomes of investors' are being distributed unequally – the highest return receives the ones with the highest incomes. Moreover, if investors attribute a large proportion of their total return on retained earnings, the part of the value they create is saved rather than spent – economic growth is stifled and opportunities for reinvestment are reduced.

Research performed by Lithuanian scientists confirms the conclusions of foreign researchers – economic inequality can have negative consequences for socio-economic progress, affecting both: the country's economic condition and various areas related to the quality of society's life:

- the level of consumption may decline as more expensive goods are more difficult to access for the poor (Deaton, 2003; Skučienė, 2008);
- barriers to the acquisition of quality housing may arise, as the accumulation of funds for the initial contribution becomes more difficult for less earning groups, getting of financial loan for housing becomes complicated as well, the payment of loans becomes harder (Skučienė, 2008);
- the need to live in overcrowded dwellings may arise, as well as problems such as no plumbing and sewerage, dripping roofs, damp walls, rotten windows or floors and dark, insufficiently bright dwellings, noise from adjacent flats, streets, industrial sites, as well as air and environmental pollution, or high crime rates in adjacent areas;
- migration may increase; according to Quinn (2006), communities with high levels of migration are characterized by high income inequality;
- the level of poverty may increase (Skučienė, 2008; Lazutka, 2003);
- It may contribute to deterioration in health status and an increase in morbidity or mortality, as income is strongly related to health status. Higher incomes provide greater opportunities to purchase goods and services that support health: better nutrition, housing, and good quality health services. In addition, research has shown that poor people are three to four times more likely to be ill than those on average or higher incomes. The risk of morbidity for people with lower incomes doubles (by eating or consuming harmful substances) (Skučienė, 2008; Lazutka, 2003).
- The level of education may decrease, as research suggests that children of poor parents are eleven times more likely to drop out of higher education programs than children of richer parents. Income inequality determines teaching quality in schools, as those with higher incomes often choose better schools (Skučienė, 2008).
- Crime rates may also rise. Braun (1997) emphasizes that in the global context, there is a tendency for income inequality to be directly related to the rate of increase in homicides (especially when it overlaps with economic discrimination based on race, religion, or ethnic groups). Therefore, it can be assumed that reducing economic inequality can reduce crime rates (Skučienė, 2008).
- Family life and the overall quality of life may deteriorate, as the deteriorating economic situation means that people work more, have less leisure time and family activities, and the number of divorces increases. On the other hand, economic inequality can be a cause of divorce as well as a consequence of divorce, as the share of single mothers raising children in society is increasing (Skučienė, 2008).

Summarizing the theories of economic inequality and different approaches, it can be stated that the origins of economic inequality are related to demographic changes, the need for food, and deprivation since the end of the 18th century.

It was mass poverty that influenced the changes in the political system and the decline in the popularity of the rich (aristocrats). At the beginning of the 19th century, economists pointed out that one small social group – the rich – appropriated most of global income. As a way to solve the problem, a steadily increasing taxation of income (the beginnings of a progressive tax system) was already proposed at that time. Market imbalance, unequal distribution of income and wealth, unlimited accumulation of capital – the main problems that have prevailed since the beginning of 19th century. Similar problems remain in today's world.

There are two important approaches to contemporary inequality theory: liberal – inequality is justified and must be; and conversely, inequality is not necessary. The first is based on the theory of the welfare economy and argues that growing inequality of income is not necessary and slows economic growth. There is also the second one, which states that growing inequality promotes economic growth.

1.2   The concept of economic inequality: Characteristics, content width, and causes

Characteristics of economic inequality. Inequality is a state of disparity or imbalance, especially in terms of social status, rights, and opportunities (UN, 2015). However, the use of the latter term in various contexts is often confusing. The concept of economic inequality is defined by its extensiveness and complexity. Economic inequality can be understood as inequality affecting the economy, or inequality of economic origin, which can be both a consequence and a cause of existing economic processes (Salverda, Nolan, & Smeeding, 2013). Economic inequality is a difference that arises when measuring economic well-being between individuals, groups, societies, and countries. Economic inequality is sometimes perceived as income inequality, wealth inequality, or the wealth gap. When focusing on inequalities, economists usually consider income, consumption, and wealth (Rakauskienė et al., 2017). Research on economic inequality is related to the essence and content of equality, equality of opportunity, and equality of outcome.

The concept of economic inequality combines several different approaches. *First, inequality is understood as an internal element of society.* According to S. Kuznets (1955), inequality did not exist only in the early history of mankind (in hunter-gatherer communities), but with the discovery of agriculture and more sedentary lifestyle, tribal leaders could enjoy a better life than the rest. Later, the pharaohs, kings, and emperors, along with other elites, also lived better than the rest of the population, and the more talented merchants or artisans lived better than the rest of the townspeople. Even in the Soviet Union, despite ideological aspirations to make everyone equal in terms of wealth, richer elite quickly formed. Therefore, economic inequality became a normal phenomenon, completely independent of the economic integration that prevailed at one time or another in human history. Thus, economic inequality has always existed whether slavery, feudal, capitalist, or socialist economies prevailed, as differences arise primarily due to objective reasons: education, skills, intelligence, physical characteristics, health status, the amount of accumulated and inherited wealth, risk, and even success.

Second, economic inequality is an organic and integral part of any society. The struggle between people is a struggle for greater social opportunities, rights, and advantages. Societies often support and even promote inequality, as it maintains the vitality of society, promotes its development, and regulates social relations (Human Development Report, UN, 2019). A society can be characterized by the manifestations of dominant inequality, with the most privileged members enjoying unlimited amount of social goods. Therefore, inequality is defined as the unequal distribution of opportunities, income, privileges, power, prestige, and influence among individuals and groups.

Third, the phenomenon of economic inequality is explained using theories of social stratification. *Inequality is a universal element of social structure and the dominant form of differentiation.* Stratification means that every society can be defined as a hierarchical ranking system. Representatives of different social strata differ in their different levels of government and material values, rights, duties, privileges, and prestige. Therefore, the hierarchical distribution of socio-cultural goods expresses the essence of social stratification, which allows society to stimulate one type of activity and interaction, while tolerating ones and inhibiting others, etc. Stratification can be defined as a hierarchy of social situations demonstrating unequal wealth, power, and the distribution of goodwill in society, through which this unequal situation is passed down from generation to generation. There is no consensus among contemporary representatives of the theory of social stratification on the criteria for dividing society into groups, but wealth, income, prestige, education, and occupation remain dominant.

Finally, the problem of economic inequality is related to the ideas of equality – equality of outcome and opportunity. Inequality of opportunities include factors that cannot be influenced by a person (e.g., ethnicity, family background, gender, etc.) and cannot be replaced by talent or effort. Meanwhile, inequality of outcome is defined by income, expenditure, and wealth. These two forms of inequality are closely interlinked and largely determine the depth of economic inequality. Reducing economic inequalities and promoting and increasing equal opportunities between different individuals is therefore crucial for the well being of members of society. To ensure equal opportunities, economic inequalities should be reduced so that people can reach their full potential.

Inequality of opportunity and outcome. To understand the nature of inequality and the multifaceted nature of the concept, it is appropriate to start the discourse with inequality of opportunity and inequality of outcome.

Opportunity inequalities include factors that are largely unaffected by ethnicity, family background, gender, etc., and which cannot be replaced by talent or effort. Inequality of outcome is usually described by income, expenses, and assets. Dabla-Noris et al. (2015) note that it is not easy to separate efforts from opportunities, especially in the context of different generations: e.g., parents' income, which they earn through their efforts, determines children's access to an appropriate education.

Researchers at Stanford University (2017) distinguish two theories – equal outcomes and equal opportunities. The theory of equal outcome states that individuals must have a certain share of resources, not just the ability to acquire them without any obstacles. An example could be literacy among school-age children: it is important for a child to become literate, not just have a theoretical opportunity to become one. However, in some cases, equality of outcomes can become a worrying factor, inhibiting individuality, and leading to the similarity of all individuals, needs, and abilities. In addition, according to Stiglitz (2012), if inequality of outcome is conditioned by the pursuit of benefits, then corruption, nepotism, improper allocation of resources, together with the accompanying negative socio-economic consequences, prevail in society. As a result, Dabla-Noris et al. (2015) argue that entrenched inequalities of outcomes can significantly impair individuals' educational or occupational choices and corresponding quality of life.

Nunez and Tartakowsky (2007) distinguish two main goals important for the development of society: (1) it is very important to reduce inequality of outcome (i.e., most often, income inequality); (2) it is very important to promote and increase equal opportunities between different individuals. The latter approach is based on the idea that all individuals should have equal opportunities to pursue their life goals, regardless of external circumstances, which the person can neither choose nor change. Proponents of equal opportunities argue that differences in economic performance partly reflect differences in individual circumstances: efforts, responsibilities, and choices. Accordingly, proponents of equal opportunities make it necessary to compare such external circumstances that determine individuals' opportunities and chances to achieve desired life goals and then to assess the inequality of outcomes arising from individuals' choices and priorities.

Despite some deviations, the theory of equal opportunities has become the reference point of the concept of justice, as confirmed by the 2005 definition of equality. The review of the World Bank's Justice and Development Report states: "We understand justice as a situation where individuals have equal opportunities to live the life they have chosen without experiencing severe deprivation" (Nunez & Tartakowsky, 2007).

According to the EU Council (2017), unequal opportunities can be one of the determinants of income inequality and vice versa. Unequal opportunities lead to greater income inequality, as the unequal starting position of individuals widens the gap between future generations' skills and earning opportunities. Conversely, if the income distribution is too unequal, the opportunities for the next generation may not be as equal, because the benefits provided by the family, which result from higher incomes and wealth, are more easily passed on to the next generation. This mutually reinforcing effect shows that policies play an important role in helping individuals to overcome social disadvantages.

According to experts of the United Nations (2015), in theory, society is given equal opportunities if certain circumstances do not lead to differences in real life. Equal opportunities exist when the consequences that people face in life do not depend on adverse random circumstances, but on circumstances for which people themselves are responsible. In this regard, it is argued that gender, ethnicity, family background, etc., do not affect any outcomes.

Inequality and human well-being. Researchers of the United Nations (2015) draw attention to a new approach to *human well-being*, its measurement, and comparison, introduced by *A. Sen* at the end of 1970s. According to Sen (2015), *"the purpose of the economy is not to accumulate wealth 'per se', but to develop people's skills; they need to become more educated, healthier and this is more important than additional buildings and roads built, and GDP deducted from it"*. Sen suggested that well-being should be defined and measured by people's actions and states (functions) and freedom of choice and action (ability). This approach emphasizes the freedom to choose one way of life over another. According to Sen theory, equalization of income is not the end goal, because not all people value income equally, and not everyone equates it with prosperity and freedom. In addition, it has been argued that this relationship is highly dependent on random circumstances, both personal and social, including a person's age, gender, family background, etc., as well as climate influence, social aspects (health and education systems, crime rates, relations in society), customs and established procedures, etc. Thus, *it is not living conditions or circumstances that should be harmonized, but real-life opportunities that give people the freedom to create the life they want* (UN, 2015).

The concept of human well-being is multifaceted. According to researchers at the United Nations (2017), well-being arises and depends on "what people have, what they can achieve with what they have, how people value what they have and what they can achieve". In other words, *well-being consists of three main dimensions*: (1) *material*, which emphasizes the importance of practical well-being and good living conditions; (2) a *relationship* that emphasizes personal and social connections; and (3) *subjective*, which emphasizes values and perceptions. These three dimensions are interrelated, and their boundaries are not yet fully established.

Despite the multifaceted nature of the concept of well-being, economists focus on inequalities in the material dimension, i.e., inequalities in income, consumption, wealth, and living conditions. There are two main approaches to this debate: the *first* concerns inequalities in outcomes in relation to the material well-being dimension, such as inequalities in income or consumption; proponents of the *second* approach examine inequality of opportunity – unequal conditions for obtaining a certain job or education.

Unequal outcomes, especially those related to income, are argued to be one of the most important criteria influencing changes in human well-being. This is confirmed by the close link between income and access to education, appropriate health care, or other services and goods. Moreover, in cases where a privileged class of society exercises its political power and influence, and where that influence affects the ability of other individuals to obtain a particular job or access to some resources, then economic inequality compromises the economic, political, and social aspects of the privileged class, limiting their ability to secure the desired prosperity.

If higher income ensures people's well-being and enable them to achieve more in life, then a person's primary income is crucial. Economic inequality (depending on its level) can both positively and negatively affect the likelihood of realizing one's potential. In other words, *to ensure at least a minimum equal opportunity, economic inequality should be mitigated so that people can realize their potential at least in substantially the same initial stages* (UNDP, 2017).

Researchers of the United Nations (2017) claim that inequality of opportunity is part of the inequality of outcomes, along with all individual circumstances such as race, gender, or ethnicity. Everything else is attributed to talent and effort (UNDP, 2017). The key difference between the two approaches is related to outcomes and opportunities. *Equality of outcomes cannot be achieved without equal opportunities, but equal opportunities cannot be guaranteed without ensuring equal conditions for people to start a full life.*

According to the EU Council (2017), unequal opportunities can be one of the determinants of income inequality and vice versa. Unequal opportunities lead to greater income inequality, as the unequal starting position of individuals widens the gap between future generations' skills and earning opportunities. Conversely, if the income distribution is too unequal, the opportunities for the next generation may not be as equal, because the benefits provided by the family, which result from higher incomes and wealth, are more easily passed on to the next generation. This mutually reinforcing effect shows that policies play an important role in helping individuals to overcome social disadvantages.

The United Nations Research Organization (2017) notes that it is still difficult to explain why the problem of unequal opportunities and outcomes continues to plague a particular section of the population. Even when monitoring and controlling criteria such as level of education, type of work, demographic variables – age, gender, and income based on race or ethnicity – the differences do not disappear. In Latin America, e.g., women still earn almost 20% less than men and this gap persist even though women have already surpassed men in their education and achievements in science. Moreover, for unknown reasons ethnic minorities (indigenous and black people) receive 13% lower salary. In addition, *persistent discrimination and social exclusion prevent people from accessing shared resources, market benefits, and public services.*

However, equal opportunities alone may not be enough to reduce inequalities in outcomes for two reasons (UNDP, 2017). *First*, high economic inequality indicates that certain processes, such as economic growth, are not fully functioning. Proponents of this perspective underestimate the importance of the structural and economic growth processes that are necessary to turn equal opportunities into equal outcomes. For opportunities to turn into outcomes, the right conditions must be in place. *Second*, despite paying special attention to the "justice" of the processes that determine material outcomes, proponents of the equal opportunity theory cannot explain why discriminatory behavior persists and why inequalities (outcomes and opportunities) do not disappear even when overt discrimination is prohibited, and the provision of basic services is ensured on a global scale.

The extensiveness of economic inequality. The content of economic inequality is characterized by specificity and extensiveness. Differentiation of income and consumption is usually the main object of most research on inequality. However, cash flow and consumption patterns are only one component of this problem. Excessive inequality, material living conditions, accumulated wealth, and its unequal distribution are also equally important indicators, but due to the problems of data collection and the complexity of measurement, the latter indicators are often circumvented. Inequality resulting from the redistribution of limited tangible and intangible resources creates the conditions for social exclusion and poverty to emerge, take root, and be constantly renewed. In terms of excessive inequality, as well as the unequal distribution of income, consumption, and wealth, poverty can lead not only to exclusion of people in terms of income and wealth but also to exclusion in social life, thus preventing life satisfaction and quality of life. Adequacy of income and consumption, material living conditions as part of economic inequality can in one case promote, in another limit human self-realization functions and creative potential, as well as promote or limit the country's economic development and quality of life.

Traditionally, economic inequality has been measured by income differentiation methods, as well as population consumption differentiation, but *to find out the real level of economic inequality, it is appropriate to analyze excess inequality and peculiarities of income inequality and concentration.* The growth of national economy does not yet guarantee an increase in social welfare, as, at the same time, there is a redistribution of national income, which, due to different understandings of social justice and socio-economic policies lead to greater or lesser inequalities in income, consumption, and wealth for the population, and concentration characteristics. Growing inequality, fostering envy among people, intensifying competition, and commercializing human relationships, denying mutual trust and empathy, reduce the quality of social life accordingly. Research on economic inequality should be given high priority in each country. Public policies should be designed accordingly to reduce economic inequalities, minimize excessive inequalities, income and wealth differentiation and tensions between members of society, and strive to ensure a dignified and high-quality life for all and the sustainable development of national economy.

Causes of economic inequality. Economists and social thinkers disagree on the causes of economic inequality. In the 19th century, the aim was to explain, justify, or criticize large differences in society. K. Marx emphasized exploitation in this context, while N. W. Senior referred to the return on capital as a reward for the capitalists' abstention and nonconsumption. Meanwhile, E. Durkheim emphasized that the causes of inequality are the need to promote the best. Meanwhile, representatives of the class conflict theory have argued that inequality is the result of a situation where people who control the values of society (power and property) can benefit themselves. According to the latter approach, the cause of inequality is the defense of government privileges, and the inequality itself is the result of influential groups seeking to preserve their status.

Representatives of the neoclassical school of economics developed a theory of marginal productivity, according to which the reward received reflects a different contribution of an individual to society (Dabla-Noris et al., 2015). In the case of exploitation, it is argued that those at the top receive what they assimilate from those at the bottom, and according to the theory of marginal productivity, those at the top receive only as much as they contribute to society. In addition, proponents of the latter approach emphasized that wage growth for workers is only possible due to the contribution of the rich, their savings, and innovation. According to the theory of marginal productivity, due to competition, all participants in the production process receive a salary equal to the marginal productivity of the individual. In this theory, higher incomes are associated with greater contributions to society. Based on this theory, the preferential tax treatment of the rich can be justified: taxing higher incomes would deprive them of a "fair reward" for their contribution to society and discourage further investment.

The ideas that justify economic inequality have lasted quite a long time because they contain some truth. Some of the super-rich made significant contributions to the welfare of society but recouped a small portion of the total contribution. However, this is only part of the search for the causes of economic inequality; other possible causes of inequality are also highlighted in the research. Inequalities in distribution can result from discrimination, exploitation, and the use of monopoly power. In addition, economic inequality is strongly influenced by many institutional and political factors, national policy on labor market regulation, taxation, money, the state budget, social benefits, income distribution, and so on.

Inequalities in income distribution cannot be explained by one economic theory, and the differences between countries are huge. Neoclassical economic theory argues that economic performance could be explained without reliance on institutes. The driving force of economic behavior is the fundamental laws of supply and demand, and it is the task of an economist to understand these basic forces. Thus, standard economic theory cannot explain how countries that are similar in application of technologies, productivity, and per capita income, etc., can thus differ in terms of the distribution of pre-tax income (Dabla-Noris et al., 2015).

Economists are still looking for ways to explain the growing inequality around the world. The global market that has opened because of globalization has encouraged workers to look for new opportunities for self-realization, higher wages, and better living conditions. Technological development has led to transformations in the structure of the economy, which have changed the demands on employees and the needs of the business (the need for middle-class employees has decreased). To a large extent, this has led to a widening of employment opportunities and the pay gap between low and high-skilled workers.

The causes of the deepening economic inequality cannot be attributed to specific factors or specific circumstances. Economic inequality is affected by a complex set of factors: from globalization-driven technological development and computerization, changes in the competitive environment and business, changing skill requirements, levels of foreign trade activity and investment to government policies, government-private balance, tolerance of the shadow economy, and corruption. However, pro-rich policies and tax avoidance are probably the main causes of economic inequality.

According to Piketty (2014), wealth and income inequality is not only the result of economics but also of politics. Stiglitz (2015) adds that inequality is the result of economic policies that favor the rich. The roots of this problem cannot be summed up in any single fragmentary factor, such as corporate tax, health, or labor market reform. They consist of a set of factors and their system. In addition, the fight against inequality requires a systemic approach, a systematic solution in many areas: financial reform, corporate governance, tax policy, antitrust, money, education, health policy, and labor law.

Reich (2010) raises similar concerns, arguing that the cause of the global crisis is not public debt, but growing economic inequality that goes beyond economic security. Reich categorically disagrees with the statement that the global financial crisis has been largely affected by people's inability to live within their means. According to Reich (2010), the problem is not that people did not live on income, but that that income did not correspond to what the growing national economy could offer. During the boom in the US economy, the middle class also expected growth in prosperity. Unfortunately, a larger share of income went to the more affluent section of society – the rich.

When the equitable distribution of income collapses, the economy needs to be rebuilt for the middle class with sufficient purchasing power (Reich, 2010). The main task of the economy is to increase the wages of most of the population (the middle section) to such an extent that it receives sufficient income to be able to fully consume and purchase the necessary goods and services. Nevertheless, it is impossible to overcome unemployment, increase budget revenues and stimulate economic growth. But, instead of new strategies that would allow the middle class to survive the crisis, political leaders, believing in the free market, opted for privatization, and reducing national influence on the economy, lowering taxes on the rich, and reforming the social security system. As a result, the wages of most citizens have been frozen, employment guarantees have decreased, and inequality has risen. Meanwhile, the results of economic growth are being used by fewer and fewer people.

Consumption is a major driver of economic growth. According to Reich (2012), it is particularly important to promote consumption in order to increase economic development (Reich, 2012). In addition, it is important to encourage the consumption of all sections of society, as an economy dependent solely on the consumption of a small group (the rich) is characterized by instability − "bubbles" and recessions in certain sectors.

In summary, it can be argued that the content of economic inequality is extensive and multifaceted, and that research on economic inequality is linked to the essence and content of equality and justice, equality of opportunity, and outcome. Reducing economic inequalities and promoting and increasing equal opportunities between different individuals is crucial for the well-being of members of society. To ensure equal opportunities, economic inequalities should be reduced so that people can reach their full potential. It is important to realize that people are not only consumers of goods and services, but also creators − not only of goods and services but also of their lives and personalities. Adequate economic well-being and quality of life are essential to maximizing a person's potential.

The content of economic inequality is inseparable from the economic condition of a country and at the same time the well-being of members of society. In this context, reciprocity is important: the better the conditions for people's self-realization and achievement of goals, the greater the potential for sustainable growth of the state economy; the more economies develop, the more opportunities there are to ensure the quality of life of members of society.

Economic inequality cannot be explained by one economic theory, just as the causes of the deepening of economic inequality cannot be summarized in specific factors or specific circumstances. Economic inequality is determined by a complex set of factors: economic, social, political, demographic, health, and psychological factors. The roots of the problem lie not in a single fragmentary factor but the combination of factors. Tackling inequality requires a systemic approach, an integrated solution in many areas, from financial and antitrust reforms to education and employment regulation.

1.3 Normal and excessive inequality

Currently, research on economic inequality pays special attention to the problem of the validity of inequality. Excessive inequality is one of the most sensitive and dramatic topics in modern society. The concepts of normal and excessive inequality are still emerging. The issue of inequality is most often addressed at the macro level. Meanwhile, there is not enough research with differentiated approaches to economic inequality and its impact on national economies. The latter approach of research is relatively new since it is quite difficult to distinguish between justified (normal) inequality and unjustified (excessive) inequality; therefore, assessment methods are not abundant, widespread, or known.

The research suggests that *economic inequality can be normal and excessive*. A certain degree of inequality is justified, depending on education, qualifications, responsibility, as well as the incentives created for innovation and entrepreneurship, development, competition, and investment promotion. However, growing economic inequality becomes a problem when it restricts people's educational or professional choices, and reduces opportunities for self-realization when their efforts are directed only at meeting the most basic needs. Excessive inequality is not just deep inequality, but one that, from a certain level of achievement, begins to interfere with socio-economic progress. Excessive inequality negatively affects economic growth, well-being, development of human resources, and increases the vulnerability of people and the state.

If normal, moderate inequality has a positive effect on national economies, then *excessive inequality is a systemic feature of economic and social distortions*; normal inequality is a positive factor, and excessive inequality is a negative factor. Inequality, depending on the circumstances in which it is implemented, can have the exact opposite effect on the social-psychological state of society. Under normal circumstances, when it does not cause a sense of economic inferiority for numerous social groups, when the majority see an opportunity to improve their situation based on their abilities and efforts, inequality positively affects the psychological state of society by encouraging constructive social forces. This is confirmed by the results of comparative studies between the United States and Europe, which show that the higher the wage inequality, the more pronounced the subjective tendency to work intensively and the higher the productivity.

Conversely, inequality, which contributes to the significant losses among the social groups, causes a feeling of helplessness, an inability to improve one's situation, and negatively affects the social-psychological state of society. In this case, inequality is not only a source of psychological tension but also distorts the social environment and motivation for economic behavior.

And yet, it is not the rich who are most affected by excessive inequality, but the middle class and the poor. The negative psychological effects of social imbalance are particularly acute in situations where unattainable examples (luxury lifestyles) are artificially advertised, causing dissatisfaction, irritability, discomfort, aggression, suicidal intentions, and wishes to protest in society, which is confirmed by statistics. The problem lies not only in the fact that these examples are highly questionable from a moral standpoint, but also, as is known in advance, they are difficult to achieve, which provokes mass frustration, especially among young people.

Inequality becomes an obstacle to the growth of human potential, not only due to the unequal distribution of goods and lack of resources for personality growth, education, health, but also due to the psychological factor. Inequality promotes a sense of unhappiness and has the most negative effect on the human psyche. Excessive inequality does not motivate a person to allocate funds and time to improve their personality, quality of life, and social status in society. This situation encourages inertia and the position of the dependent, increases the national budget expenditures, and, at the same time, reduces the economic contribution of labor activities. All these factors hinder the country's economic development.

The negative social consequences of excessive inequality are threatening because of "poverty trap": people realize that they cannot break free, there are no strong enough socio-economic incentives, and people lose their will to be active. A person gives up, resulting in an increase in the number of suicides, an increase in the number of cardiovascular diseases, an increase in crime, etc.

Excessive inequality can lead to policies that are detrimental to economic growth. Inequality not only affects the drivers of economic growth, but it can also lead to inefficient public policies. For example, it can provoke a hostile reaction to growth-friendly liberalization and protectionism against globalization and market-oriented reforms. At the same time, stronger power could lead to more limited productivity and growth-enhancing public goods that benefit the poor.

Excessive inequality hinders poverty reduction. Economic growth is less effective in reducing poverty in countries with high levels of inequality or where the results of economic growth are not distributed in favor of the poor. High inequality means that a larger portion of the population is more vulnerable to poverty.

According to the research of the World Bank, inequality becomes excessive from 30% to 40% of the Gini coefficient level. Excessive inequality is not just deep inequality (deep does not mean excessive), but one that, from a certain level, does not stimulate but stifles economic growth and has negative socio-economic consequences. In this sense, inequality is evolving from a social problem to a major impediment to economic growth and quality of life.

In 2017, the European Commission pointed out that "to some extent, inequality can lead to investment in human capital, mobility and innovation", but "growth can be jeopardized if inequality becomes excessive". However, scientific substantiation, namely when inequality becomes too great (excessive), has not been done, and the research on this topic is more of a theoretical hypothetical nature.

The EU Council (2017) emphasizes that *inequalities of some magnitude can stimulate investment in human capital, mobility, and innovation*. Economic incentives (which are important for growth) are effective if a person can achieve better results through their hard work. However, *if inequality becomes excessive, growth could be jeopardized.* This is especially true in cases where inequality is caused by increased poverty in the lowest income distribution segment. If those on the lowest incomes (or those with the least wealth) lack the resources to invest in their skills and education, they may not be able to reach their full potential, to the detriment of overall economic growth.

A study by the IMF entitled "Causes and Consequences of Income Inequality: A Global Perspective" (2015) states that growing inequality is becoming an increasingly acute problem. Inequalities in developed, emerging markets, and developing countries have increased and are receiving considerable attention. A recent study by the Pew Research Center (PRC) found that more than 60% of respondents around the world see the gap between rich and poor as a major challenge. Indeed, the PRC study found that while education and hard work were seen as important factors in improving their social status, dating the right people, and belonging to a wealthy family were also seen as important factors. The study, therefore, highlights significant potential barriers to social mobility. For these reasons, the extent of inequality, its drivers, and the questions of possible actions become one of the most important issues under discussion among politicians and scholars.

IMF experts (2015), in response to why the issue of inequality is so important, argue that equality, like justice, is an important value in most societies. People, regardless of ideology, culture, or religion, care about inequality. Inequality can be a signal of income mobility and lack of opportunities – a reflection of persistent deprivation in certain sections of society. The growing income gap also has significant consequences for economic growth and macroeconomic stability; it can lead to a concentration of political and decision-making power in the hands of a few, cause inefficient use of human resources, political and economic instability unfavorable to investment, and increase crisis risk. The economic and social consequences of the global financial crisis and the obstacles to global economic growth are what prompted to examine growing income inequality more closely.

IMF experts (2015) emphasize that, first, economic policymakers must pay more attention to the poor and the middle class for two reasons. *First,* a more even distribution of income, i.e., increasing the share of income allocated to the poor and maintaining the income level of the middle class has a positive effect on economic growth. *Second,* inequality develops in different directions in developed and lagging countries.

Discussions about inequality often distinguish between *inequality of outcome* (measured in income, wealth, or expenditure) and *inequality of opportunity,* which includes factors beyond a person's control, such as gender, nationality, place of birth, or family structure. Inequality of outcome is determined by differences in opportunities and a combination of individual abilities. At the same time, it is not easy to separate abilities from opportunities, especially in different generations. For example, parents' income, which they earned through their efforts and abilities, determines their children's access to education. Rawls (1971) argued that the distribution of opportunities and outcomes is equally important and informative for understanding the nature and extent of inequality worldwide (IMF, 2015).

Asking whether inequality really should be viewed negatively, IMF experts (2015) argue that a certain degree of inequality can be justified if inequality encourages improvement, competition, saving, and investment for progress. For example, higher education and wage differentiation can foster human capital accumulation and economic growth despite higher income inequality. Inequalities can also have a positive impact on growth through incentives for innovation and entrepreneurship and, especially in developing countries, by enabling at least some individuals to gain the minimum necessary to start a business or education (Barro, 2000).

When does increasing inequality become a problem? High and persistent inequalities, especially inequalities of opportunity, can increase social costs. Inherent inequalities in outcomes can significantly reduce individuals' access to education and career choices. Moreover, inequality of outcomes does not create "right" incentives if it depends on benefits (Stiglitz, 2012). In this case, individuals are encouraged to focus their efforts on favorable behavior and protection, which in turn leads to misallocation of resources, corruption, and nepotism, together with the negative socio-economic consequences of these phenomena. Citizens may lose confidence in institutions, leading to a loss of social cohesion and faith in the future.

Inequality affects the drivers of economic growth. Why are growing income disparities affecting economic growth? Higher inequality reduces growth by depriving lower-income households of the opportunity to maintain their health and accumulate material and human capital. For example, it may lead to underinvestment in education, as children from poorer families end up in lower-quality schools and have fewer opportunities to access higher education. For this reason, labor productivity could be lower than it would have been in an environment of greater equality (Stiglitz, 2012). Corak (2013) similarly concludes that in countries with higher income inequality, a lower degree of intergenerational mobility is also generally observed, with parental income being a more important indicator of children's income. Increasing income concentration could also reduce demand as the wealthy spend a smaller share of their income than the middle class and low-income earners.

IMF experts (2015) argue that inequality inhibits investment, as well as economic growth, by contributing to economic, financial, and political instability (financial crises and conflicts).

- *Financial crises.* There is growing evidence that there is a causal link between the growing influence of the rich and the stagnation and crises of the poor and the middle class, while at the same time causing direct damage to short and long-term growth. Various studies suggest that the long period of increased inequality in developed countries is linked to the global financial crisis due to excess in loans and less stringent lending standards, and opportunities for lobbyists to deregulate financial activities.

- *Global imbalance.* Higher incomes of the richest, combined with liberalization of the financial sector, which could be a response to growing
  income inequality, are associated with significantly higher external deficits. Such large imbalances can pose a challenge to macroeconomics and/
  or financial stability, and therefore to growth.
- *Conflicts.* Excessive inequality can undermine confidence and social
  cohesion and is therefore also linked to conflicts that prevent investment.
  In summary, inequality can intensify the frustration of some groups or
  reduce the alternative costs of initiating or participating in conflict.

In summary, a differentiated approach to economic inequality emerges, i.e.,
inequality can be justified – normal and unjustified – excessive. A certain
degree of inequality can be justified if development, competition, and investment in the public well-being are encouraged. In the meantime, inequality becomes unjustified when it no longer stimulates, but, on the contrary,
hinders economic growth and has negative socio-economic consequences.
Excessive inequality is becoming a systemic feature of economic and social
distortions, holding back national economic growth, and adversely affecting
people's quality of life.

Considering the arguments mentioned above, a scientific problem arises
which is caused by three main circumstances: first, there is a lack of methods in the scientific literature to distinguish between normal inequality and
excessive inequality; second, empirical research to objectively assess the
impact of increasing economic inequality on socio-economic progress is
insufficient. These circumstances allow formulating a scientific problem and
relevant questions: what inequality is, how reasonable, justified, i.e., normal
it is, and when it becomes unreasonable and unjustified, i.e., excessive, and
how it affects socio-economic progress.

1.4 Wealth inequality – the major indicator of excessive inequality

Over the past few decades, global economic inequality has increased on
a surprising scale, while *wealth inequality has been growing faster than income
inequality*. Notwithstanding the fact that, after the global financial crisis of
2008, the income of the wealthiest 1% of the world's population decreased;
however, they have at their disposal 20% of the worldwide income, while
the wealthiest 1% of the planet's population, whose the total value of assets
amounts to more than $1 million, control 45% of all the wealth of the earth.
The latter fact alone confirms that *wealth inequality, rather than income inequality, highlights the depth of economic inequality.*

Notwithstanding the fact that the rapid economic growth in some countries (e.g., in Asia, especially in China and India) has decreased the number
of people living in absolute poverty; however, the economic gap continues to deepen because the wealthiest population of the world derives the
most benefits from economic growth and the assets accumulated by them

are exponentially increasing. Alvaredo, Chancel, Piketty, Saez, and Zucman (2018) emphasizes that *the wealthiest 1% of the world's population receives twice the benefits from economic growth, which is received by the remaining 50% of the poorest planet's population.*

As an integral part of economic inequality, wealth inequality is an acute problem all over the world, and this is confirmed by the following facts:

- The rich, whose total value of the assets is higher than $30 million, own 11.3% of global assets; however, the latter super-rich represent just a small part of (0.003%) of the planet's population. Furthermore, the assets of the latter grew the fastest: in 2016, their assets increased by as much as 25.5% compared to 2017.
- The well-off of the world's population, whose total value of their assets is estimated at over $100 thousand, represents less than 10% of the global population; however, they dispose of 84% of global assets.
- Whereas the people, whose total value of their assets is less than $10,000, represent 64% of the world's population.
- During the period from 2009 to 2017, a number of the billionaires, whose property was estimated at 50% of the assets of the world's poorest population, dropped from 380 to 42, i.e., 42 persons control assets of the same value as the poorest 50% of the planet's population.
- Wealth inequality is especially significant in the United States since even 42.5% of national wealth is controlled by 1% of the richest society representatives. Besides, the United States differs from other countries by the fact that 5.3 million of their population have accumulated at least $ 1 million of financial assets (not including real estate assets or consumer goods).
- The majority of the world's billionaires (over 65%) live in Europe or North America (41% in the United States). There are 70,500 people residing in the United States, whose value of their assets amounts to $50 million, which is more than double the value of the assets owned by the population of the other five richest countries, which amounts to $50 million, i.e., China (16.5 thousand), Germany (6.3 thousand), Great Britain (4.7 thousand), Japan (3.6 thousand), and France (3 thousand).
- The percentage of billionaires worldwide is rapidly increasing in China, and over a year from 2017 to 2018, their percentage increased from 5% up to 7%. Moreover, in China, there is an extremely rapid growth of persons, whose assets are valued at $50 million: in 2017, the latter amounted to 9,600, and in 2018 – to 16,500 (Global Wealth Databook, 2018).
- An obvious reflection of wealth inequality is the Forbes 400 list of the super-rich. The total value of assets owned by three persons top-rated by the magazine in 2018, i.e., Jeff Bezos, founder of Company Amazon, Bill Gates, founder of Corporation Microsoft, and investor Warren Buffett, was higher than half (50%) of the poorest population in the United States.

- Since 1982, three dynastic families in the United States – the Waltons, Kochs, and Mars – have increased the value of their wealth by nearly 6,000%. Meanwhile, the wealth of an average household in the United States has reduced by 3% over the same period. The latter three dynasties have accrued assets for $348.7 billion, which is 4 million times more than the assets accrued by an average family in the United States.

The emphasis on the problem of wealth inequality and the search for ways to reduce inequality are not aimed at equalizing wealth and income. The main aim of highlighting this issue is the equitable distribution of the benefits of economic growth and increasing prosperity among members of society; proportionate remuneration for work performed and contribution to the well-being of the community; fair and equal opportunities to accumulate and dispose of such capital or wealth as is necessary for a person to feel a safe and full member of society; and everyone has the same potential to choose a life that meets their needs.

It should be noted that when wealth is distributed unequally throughout society and the higher wealth exclusion is among members of society, the less the rich want to invest in the creation of community welfare (education, medical care, personal security, etc.). Thus, society destabilizes, and a widening gap between the richest and the rest of society and increasing inequality destroy confidence in the state, society, and self. Confidence of the richest members of society in their status, government and possible impact is illusory and temporary: there is no apex of a pyramid; the rich cannot fully feel secure if the rest of society is weak. The poor part of society has a huge power which can at any time outbreak in different forms as economic inequality grows – strikes, protests, disturbances, explosions in crimes, conflicts among different social groups, uprisings, etc., what actually destabilizes society and is a threat to the whole state. According to Lazutka's (2018) opinion, research has shown that increasing inequality causes a lot of social issues. For example, the United States has much greater economic inequality than in Europe, which is associated with the problems arising in the United States: there are many more mentally ill people, crimes, underage pregnancies, and drug use than in Europe. All those issues are associated with economic inequality. The research conducted by Pickett and Wilkinson (2018) demonstrated that society in the countries, where great wealth inequality is, suffers more from various social problems: people are more overweighed, they have a higher incidence of mental illnesses, higher rates of murders and suicides, more under-age girls have children, a larger number of prisoners, a higher level of bullying among children, increased drug use, etc., than in the countries with more uniform distribution of wealth.

In this context, Piketty (2017) states that it is wealth inequality is one of the key drivers of terrorism when oil revenues started concentrating in only a few states with a relatively small number of people: within the region between Egypt and Iran, including Syria, some monarchies where only 10% people of this huge region live and control approximately 60–70% of assets.

Economic globalization provides maximum benefit to the rich people. For example, after the economic crisis of 2008, notwithstanding the fact that all groups of society lost a large portion of their savings and assets, wealth inequality worldwide did not reduce but even increased. While states were looking for ways how to escape from economic recession, the richest planet's people continued to increase rapidly their wealth. According to Stiglitz (2015), *inequality exists because 1% of the national population aims at that.* Fiscal policies are exactly the proof of that: graduating tax relief on taxable capital gain facilitating the accumulation of assets has created particularly favorable conditions for certain groups of society, and the extent of present-day inequality is mostly determined by the absence of antitrust and manipulation of the financial system. According to Lazutka (2018), the rich have a lot of possibilities of affecting legislators and do that in different ways, both legally and not very legally. To change the situation, the richest population of the country must pay more taxes in order to contribute to the common good.

The absence of the leverages of uniform distribution and concentration of assets and cash in one small group (the richest) of society lead to the appearance of economic dysfunction because they reduce consumption. People who receive higher income spend comparatively a small part of their income compared to the persons receiving lower income. It is confirmed by calculations made on the basis of the assets and earnings accrued by M. Romney, politician and businessman. On the basis of the data of 2010, M. Romney's income was $21.7 million; even if this person decided to increase his spending, manually he would spend a rather small amount of his income in order to support his family and a few houses owned by him. However, Stiglitz (2015) notices that if the same amount of money were divided among 500 persons, e.g., by providing employment for each of them, paying $43 thousand per year, it is obvious that most of his money would be spent. Therefore, an assumption is worth highlighting that *the more significant differentiation is, the less consumption and the general demand are.* With the onset of the latter trend, the general economic demand becomes less than supply, which typically means an increase in unemployment, which reduces demand even further.

The "bubble" of high-tech companies developed in the last decade of the 19th century and a real estate "bubble" of the first decade of the 21st century slightly slowed down the growth of differentiation between wealth and income. However, an answer to the question, what can change and halt the increase in economic inequality, is yet to be received.

According to Piketty (2015), tax issues in the global world is the future issue of the whole capitalist system (but not policies of separate states). Taking into account a huge and still growing scale of wealth inequality, Piketty urges to resolve this problem as soon as possible and to do that by using globally the only measure, i.e., by introducing a 10% wealth tax to all rich people of the world. It is not suggested to introduce a global tax on wealth abruptly but to agree on a clear plan and to increase it annually at least by a few tenths of a percentage unit, and thus to reach 10%. The key essence of the measure suggested is that it must be implemented globally and applied to all rich people without any exceptions and loopholes. In this respect, states must mobilize all together because, otherwise, if some countries start increasing taxes, while others decide to decrease them, the latter will raise their capital in this way by attracting it. Cyprus and Luxembourg, which collect incomparably more wealth taxes than other countries where this tax is higher, stand as evidence. The latter countries are referenced as "tax heavens" in the European Union (EU) and the latter provide all conditions for the existence and growth of welfare.

Large international companies and natural persons exercising tax advantages, different legislative gaps and "tax heavens", escape huge amounts of money. According to the EC, such tax evaders do not annually pay as much as EUR 1.1 trillion of taxes in the EU. In this respect, the EU has brought several cases to court against international corporations such as Microsoft, Apple, Google, etc.

Nonprofit organization Oxfam dealing with poverty and deprivation also emphasizes that it is necessary to aim to withdraw globally the conditions for so prevalent evasion of tax on capital and to curb so the growth of wealth inequality. Thus, more funding for education and health protection would be accrued what is especially important in combating inequality. Moreover, the IMF also notes that countries should increase taxes on capital and assets, because growing wealth inequality is already threatening the world with serious social disturbances – protests of the Arab Spring, the Maidan protest revolution in Ukraine, etc., is the consequence of huge weal inequality.

Economic inequality is a serious problem of the whole world, and the extent of this problem is mostly reflected by wealth inequality. The gap between the richest and the rest of society continues widening as the richest planet's population receive the major benefit from the growth of economics, and the assets accumulated by them is rapidly increasing. Hence, inequality is the result of pursued economic policies. The concentration of tangible assets in the hands of one group not only destroys the principles of social justice, but it also restricts the opportunities of the members of the rest of society and the quality of their life, and simultaneously suppresses and discourages the socio-economic progress of the state itself.

1.5 A modern perception of socio-economic progress

Society's perception of progress has been changing depending on the priorities and goals of mankind. The conception of socio-economic progress has evolved from the term of economic progress, to which particular attention was paid by the scientists such as Hansen (1939), Ayres (1944), Fagan (1935),

Clark (1940), Davis (1947), and other research in the inter-war period of the last century.

The economic progress defined by Hansen (1939) combined discoveries, the development of territories, the discovery of new resources, and population growth. The latter criteria can be assumed to be productivity growth supported by an increase in resources and developed production. Ayres (1944) has also treated economic progress as the growth of production volume based on productivity by suggesting that the benefit of such growth will be distributed for everyone in the longer term. Fagan (1935) has defined economic progress slightly wider by combining an increase in the use of labor and natural resources (i.e., the growth of productivity) together with the distribution of assets received. Clark (1940) has identified economic progress with the growth of productivity; however, he emphasized "a fair distribution of fruit": by increasing disposable income, providing the conditions for appropriate leisure time, etc.

Davis (1947) is one of the first scientists who have adopted social and sustainability dimensions to economic progress, and he has highlighted in his studies: (1) the proportionality of "the fruit" of productivity growth for different interested groups; however, he has also emphasized that (2) the significance of the repeated use of the resources, which were employed for the growth of productivity, as well as has studied a lot of other social and environmental effects.

Productivity growth is the driver of economic progress but it is not the only component. According to Davis (1947), economic progress is driven by productivity growth including:

- Balanced distribution of the assets developed by productivity growth;
- Reuse of the resources, which were employed for productivity growth, which is required to increase production volumes;
- Collection of initial capital for the development of the future production generating added value;
- Setting of minimal social expenditures, which are necessary in the cases of unemployment, accidents at work, or illnesses.

Every above-mentioned component is necessary so that productivity may grow to the overall social and economic progress. Thus, Davis (1947) has expanded economic progress up to socio-economic progress.

Based on the ideas of Davis, Grifell-Tatje, Lovell, and Turon (2018) also state that adding value through *boosting productivity* is a necessary requirement for socio-economic progress; no value can be added without productivity growth. However, the business contributes to socio-economic progress only if value added is *distributed with balance* for all relevant interested groups. The need to reuse resources (especially in order to save them) is equally important; however, this can be done only because of improved technologies, i.e., *technological progress*, fostering productivity growth. Grifell-Tatje, Lovell, and

Turon (2018) have enhanced the social dimension influencing productivity growth and have introduced the criteria of social costs and social benefits exactly highlighting *adding social value* and a balanced distribution of social value to the whole society as to the interested group.

Thus, according to Grifell-Tatje, Lovell, and Turon (2018), business productivity growth generates value-added, which is necessary for economic progress, which also requires a balanced distribution of generated value-added, the resources, which have been changed by productivity growth, repeated employment, and development of a positive social effect. The concept of equality (inequality) is incorporated in all these components of this socio-economic progress; therefore, the contribution to socio-economic progress tends to reduce inequality.

Socio-economic progress in the contemporary context. It is emphasized that the peculiarities of the conception of economic progress highlighted by Fagan, Clark, and Davis have been ignored for almost 70 years, and they obtained some exclusive attention after highlighting them by Stiglitz, Sen, and Fitoussi and the significance of socio-economic progress by the OECD in the modern context.

The processes of globalization (transformation processes) have prompted to review evaluation indicators of the economic situation. For a long time, the gross domestic product (GDP) has been treated as the core indicators of economic efficiency and activity, and it also reflects the quality of life. However, at the beginning of 2008, by the initiative of N. Sarkozy, President of the Republic of France, the Commission for Assessment of Economic Condition and Social Development – *"Commission of Happiness"*, which included the best-known economic leading figures such as Stiglitz, Sen, and Fitoussi, was established. The Commission had an ambitious goal – *to assess the limits of decency of the most popular macroeconomic indicator, i.e., GDP, and to prepare a system of alternative indicators reflecting the current welfare of society and its sustainability.*

Stiglitz, Sen, and Fitoussi (2010) have noted that as an indicator GDP does not completely reflect the economic situation, welfare, and progress made of a country. For example, if GDP of a country grows, simultaneously economic inequality can also rapidly increase and the quality of life of the population can worsen, despite the fact that the average income increases. It often happens because of the unequal distribution of income or wealth concentration in a group of certain persons, and only this group is allowed to exercise and enjoy the advantages of increasing income or growing assets, while the rest group of the population does not feel any significant benefit. Even traffic congestions within metropolitan areas can affect the growth of GDP due to higher consumption of fuel, but air pollution will not really lead to the improvement of the quality of life.

GDP indicator is not ideal not least because it does not warn about the upcoming economic crisis and the risks to the financial system. In 2004–2007, most countries of the world, including Lithuania as well, could be proud of impressive rates of GDP growth, which very fast became a dramatic decline. "Bubbles" of prices and profit, to the explosion of which neither the government nor the public was properly prepared, resulted in the financial and real estate markets. In other words, there is a need arising to develop a system of early warning indicators, which enables the government to take timely actions and to ensure the sustainability of socio-economic processes.

A report prepared by Stiglitz, Sen, and Fitoussi (2010) reveals and highlights the difference between the indicators reflecting the current welfare and sustainability. To analyze the above-mentioned problems deeper, *"the Commission of Happiness"* has organized itself into three working groups for *(1) Classical issues of GDP (economic), (2) Quality of life, and (3) Sustainability, and so enhanced that they were the dimensions of the essential parts of socio-economic progress.* As a consequence, they prepared a document, which recommended applying the system of indicators reflecting the current and future generation welfare wider instead of the statistics oriented to production. The most significant conclusion of the group, which had analyzed the system of indicators of quality of life, was that *it was necessary to move its center of gravity from the measurement of manufacture and production to the statistics of people's welfare and quality if their life.* Country's GDP is an indicator accepted to describe economic activity; however, it is considerably less suitable for establishing of changes in living standards. Besides, the current situation and sustainability cannot do without each other – today's well-being can be achieved by destroying natural and other resources, polluting the atmosphere, and vice versa harmonious and moderate economic development ensures guarantees for long-term progress. That's why, in the opinion of the members of the Commission, it is necessary to devote considerably more attention than so far to such broadly understandable and relatively easy quantitatively measurable indicators as the concentration of gas causing the "greenhouse" effect in the atmosphere and its approach to the critical level.

Thus, *the main conclusion of the report prepared by "the Commission of Happiness"* emphasizes that *such general economic indicators as GDP, inflation, budget deficits, are far from reflecting the real economic situation in the country as the real economic condition is shown by the indicators of quality of individuals' life. Therefore, assessment of the economic situation of the country requires moving the center of gravity to the criteria of the welfare and quality of people's life.* At the same time, "the Commission of Happiness" emphasized that it was not possible to reflect the welfare of modern society with one or two indicators, and, consequently, it offered a few dimensions of its assessment:

- Material indicators of the measurement of living standards (income, consumption, and wealth);
- Living environment and ecology;
- Education;
- Health,

- Job and personal activities;
- Social contacts and participation in the community;
- Political "voice" and opportunities of influencing government decisions; and
- Physical and economic security.

After the insights produced by *"the Commission of Happiness"* about the advantages and disadvantages of GDP as a macroeconomic indicator, and recommendations provided for new methods and indicators offered for the measurement of socio-economic progress, further scientific and political discussions were based on all the provided arguments and that supported. For example, in 2016, in Davos, economists and scientists unanimously agreed that GDP is the way, which is no longer suitable to assess economic "health", and it is a necessity to find a new method for measurement. Lagarde, Stiglits, and Brynjolfsson (2016) emphasized that *the world is changing; therefore, the methods how to assess progress must also change simultaneously.* As Brynjolfsson (2016) states, it is necessary to find a new model of economic growth; as it is required to review the business (checking the correctness of its direction) depending on the environment changed, so the methods for measurement of economics must also be audited.

Stiglitz (2016) notes that GDP in the United States grew annually (except for 2009); however, most Americans live less well than the one-third of a century ago. The benefits of economic growth have reached the top group of society, while the salaries and wages of the rest of society at today's value are lower than they were 60 years ago. Thus, the current economic system does not serve the majority of society. Within this context, JStiglitz (2016) emphasizes: "That what we measure shows what we do; and if we measure wrong things, it means that we also do wrong things". Therefore, it becomes obvious that GDP is no longer a measurement, which fully estimates economic situation, quality of life, assesses the progress in these areas, and simultaneously combines economic and social dimensions. In this way, the significance of the conception of socio-economic progress combining the above-mentioned attitudes arises.

The International Centre of Public, Social and Cooperative Economic Research and Information emphasizes (2007) that the socio-economic conception is closely related to the conceptions of progress and social cohesion. The contribution by the social economy to the European society is significantly larger than the one, which is reflected by GDP from an economic point of view. The potential of this economic sector to generate social value-added is enormous, which is the same as the multifaceted and especially qualitative-nature realization of this potential; therefore, it is not always easy to understand it and to evaluate it quantitatively (CIRIEC, 2007).

It should be emphasized that the progress of the social economy can be observed even within the period of an economic and social crisis. Actually, the social economy is an example of resilience, and it continues to develop

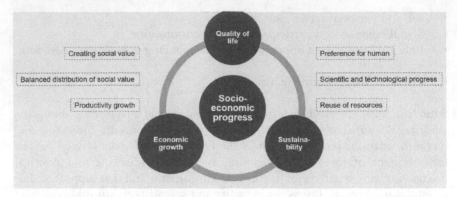

Figure 1.1 Conception of socio-economic progress.

Source: Made by authors.

when other economic sectors are struggling to recover. The conception of the social economy emphasizes that we need the economy combining social, economic and financial dimensions, which can create wealth and be evaluated not only in accordance with its financial capital, but also social one, which probably is preferred. Development, double-digit profitability, and profit are not the core goals; first of all, it is pursued to contribute to the global interest, social cohesion, and the well-being of society.

In summary, it could be argued that it is increasingly acknowledged that basing on the GDP as the only guiding force does not provide sufficient information on how the economy operates in respect of citizens, how it influences the quality of their life, and what long-term affect it has on sustainability. In this context, the significance of socio-economic progress increases.

Socio-economic progress is a complex notion and a difficult process, which strives to ensure welfare to the current and future generations. The conception of socio-economic progress of (Stiglitz, Sen, & Fitoussi 2010) has been distinguished. It combines three components, i.e., economic growth, quality of life and sustainability (see Figure 1.1) and thus a requirement to a systematic and complex assessment of modern society is presumed.

On the other hand, although the significance of sustainability is indisputable, however, this dimension is an integral part of present-day economic growth and quality of life. Sustainable and moderate development of the economy only can ensure guarantees for long-term progress to the existing and future generations and quality of their life.

According to Stiglitz, Sen, and Fitoussi (2010), the welfare of society is affected by economic factors (e.g., income) together with noneconomic indicators of individual's life (e.g., it is important what individual's occupation is, how he realizes himself, how he feels, what environment he lives in, etc.). However, with the development of modern economies, it is very important not only to evaluate (measure) the condition of country's economy, but also to assess the quality of people's lives, and all measurements should be in the context of sustainability.

Therefore, *economic growth and quality of life are the key components of socio-economic progress. These two dimensions of socio-economic progress, i.e., economic growth and quality of life, combine two attitudes – quantitative and qualitative.* Economic growth is mostly expressed and evaluated with quantitative indicators – GDP growth, unemployment rate, labor productivity, investments, etc. Meanwhile, quality of life is often reflected by qualitative indicators – health, education, economic and physical safety, living environment, general experience of life, etc.

Socio-economic progress requires the involvement of society when developing social policies and economic initiatives. The main *goal of socio-economic progress is continual improvement of the welfare of an individual, groups, a family, a community, and the entire society both in terms of economy and quality of life.* This means a long-term improvement of the economy and quality of life of country's people, which usually is reached by reducing economic inequality, social exclusion, and poverty. Sultanoglu (2004), authorized representative of the United Nations Development Programme in Lithuania, emphasizes that socio-economic progress has a positive impact on the social development of the country's people; however, it is especially *important to ensure that all citizens can use socio-economic progress by* creating their own well-rounded lives.

Dudin, Ljasnikov, Kuznecov, and Fedorova (2013) emphasize the significance of innovative transformations, i.e., progress of science and technologies and innovations, to reach socio-economic progress. The basis for socio-economic progress is knowledge applied as a strategic source of development and changes. Namely, knowledge expressed as unique ideas and solutions can help to achieve the best satisfaction of individual or social needs, as well as economic growth of states.

According to the Commission of the European Communities (2005), the growth itself cannot be the principal goal. Enhancement of competitiveness and giving a new impetus for growth are not the goals but measures to achieve it: the growth itself does not guarantee social solidarity and sustainability. That is why solidarity and security together with welfare must continue to be the main goals of Europe. Besides, such policies, which ensure that different goals strengthen each other, must be pursued. Actions, which foster competitiveness, growth and employment, economic and social cohesion, and a healthy environment, strengthen each other. Prosperity will not fully be realized without solidarity and cohesion: the economy is there to serve people, not the other way round.

Significance of quality of life in the assessment of progress. The concept of quality of life and its meaning have changed historically. For a long time, it has been believed that GDP growth and as high personal income as possible are a guarantee for the quality of public life. Finally, research and history have proved that GDP, which was considered as the key measure of national economic development and quality of life for a long time, can grow for years;

however, it cannot ensure the quality of life for members of society because of unbalanced economic and social policies.

Quality of life is an object of interdisciplinary studies and a multidimensional indicator reflecting the efficiency of the economic policies pursued in the state and the level of well-being in society. Quality of life is the entirety reflecting the physical and mental health status, evolution of population and welfare of the family, social and physical security, quality of a living environment, the relation between professional activities and leisure time, satisfaction level of material, and cultural and spiritual needs. And it is assessed at the macro-prudential level (countrywide) and micro-prudential level (from an economic point of view). Consequently, quality of life can be assessed in accordance with objective social and economic facts (objective indicators of quality of life), as well as in accordance with subjective assessments of these facts by individuals (subjective indicators of quality of life) (Gruževskis & Orlova, 2012).

Quality of life can be defined not as the state of individual satisfaction of members of society but as existence of the conditions necessary for happiness in society. The necessary conditions are related not to person's desires and expectations but rather to an opportunity to satisfy the needs. Immediate needs are the same for all representatives of mankind, regardless of the region where people live. An important characteristic of common needs is that they are defined, while people's desires and expectations have a tendency for continuous change and growth (Gruževskis, 2012). The needs of the population and attitude toward them in many cases also depend on the country's history, economic development, mindset of society, education, and cultural level. Economic growth is no guarantee of an increase in social welfare because simultaneously national income is redistributed, and because of a different understanding of social justice and economic policies pursued in a state, such redistribution causes higher or lower inequalities of income and consumption expenditure of the population. The consequence of particularly unequal distribution of income and expenditure is population polarization, high indicators of differentiation, and rates of poverty. Due to increasing inequalities inspiring envy among people, growing competition, and commercialization of relations among people what destroys confidence and empathy, the quality of social life also worsens.

GDP is a quantitative but not qualitative indicator; therefore, it is not able to show qualitative changes, and, consequently, the absolute improvement of quality of life (Starkauskienė, 2011). Accordingly, economic growth expressed as GDP, improvement of macroeconomic indicators (government deficit, public debt, and decrease in inflation) cannot be identified with the equivalent improvement of quality of life. Therefore, besides economic indicators such as employment and unemployment rates, rates and distribution of income and consumption, and housing and living environment, the crucial role is played by noneconomic indicators and those of sustainable economic prosperity, which are orientated toward psychologic capital, happiness, health, safety, social relations, social contacts, cultural and spiritual treasures, conservation of nature and healthy environment, sustainable production, and sustainable consumption (Rakauskienė et al., 2015).

In summary, it could be argued that a lot of authors in their morphology refer to the provision of Stiglitz (2009) that economic progress cannot be separated from social progress; both these areas must be integrated into one conception and named as socio-economic progress. The concept of socio-economic progress is based on the fact that the indicators of economic activities such as GDP and per capita GDP are not sufficiently expressive in order to define country's socio-economic progress. Therefore, the conception of socio-economic progress must include the indicators of quality of life and sustainability.

Thus, the key factor promoting socio-economic progress is the quality of life of the members of society. It has largely been influenced by a changing paradigm of the economy when universal monetary methods based on the weights of monetary and fiscal policies are replaced with unorthodox conceptions of the economic development directed toward the interests of everyone in society and assurance of the quality of their life. Consequently, although for decades an approach that economic growth was the fundamental of socio-economic development had prevailed, recently, the emphasis is placed on the indicators of the quality of people's life.

Economic inequality and economic growth. In the middle of the 20th century, little significance was attached to the inequalities of income distribution because it was believed that, on the basis of the hypothesis driven by Kuznets, that income increased steadily and economic inequality would finally start to reduce. However, over time, the number of studies carried out which showed that the opposite was the case, i.e., *economic inequality had a negative impact on economic growth and welfare of society*, had been increasing.

Economic inequality restricts socio-economic welfare. Income and consumption inequality, redundant inequality, and unequal distribution of assets not only rebut the principles of social justice, but also decrease the opportunities of accessing education, parenting, protection of health, culture, qualitative housing, and living environment to the members of society. Sufficiency of income and consumption, possession, disposal, or control of assets lead to individual's material and moral safety, self-confidence, self-esteem, and quality of life. Meanwhile, unequal distribution of income, consumption and assets may have an impact on the growth of people's polarization, high indicators of differentiation, and corresponding poverty rates.

A question about the interaction of economic inequality with economic growth is increasingly brought up in order to understand a possible impact on the level of development of states and competitiveness, and thus on the quality of life of the members of society. Transformation processes in the economy had a large impact on the situation when universal monetary methods based on the weights of monetary and fiscal policies are replaced with unorthodox conceptions of the economic development directed toward the interests of everyone in society and assurance of the quality of their life. Prioritization of the public interests influenced the economic growth and public satisfaction with their life of the leading EU Member States such as Germany, France, Austria, Scandinavian countries, etc. In turn, other states devaluating human resources maintain their position that the radically liberal doctrine of the Washington Consensus will improve the level of public life and strengthen the competitiveness of the state, and ensure socio-economic progress, although it is increasingly being criticized.

It should be noted that it is very important to create such an environment where economic growth is promoted but not slow down. Increasing economic inequality has significant implications for economic growth and macroeconomic stability; it might be the reason for the concentration of political and decision-making power in the hands of a few, determine the suboptimal use of human resources, cause political and economic instability unfavorable for investment, and increase the risk of crises (Dabla-Noris et al., 2015). The need to understand complex relationships between economic inequality and economic efficiency, economic inequality, and welfare of society is really a sufficient reason for the appearance of economic differentiation on the top of the list of problems addressed by economists because the latter phenomenon is the field where the insights of economists will particularly help studies related to this issue and searches for the ways to reduce inequality and their realization in practice.

Economic inequality and sustainability. The EU will strive for becoming a climate-neutral society by 2050, when basing the welfare of the future generations on the neutralized impact of greenhouse gas, the economy currently poses huge challenges already. The transformation of modern society into a climate-neutral society is an urgent challenge in order to ensure currently the opportunities of welfare for future generations by balancing economic, technologic, ecologic social, and psychologic development. This goal is the foundation of the European Green Deal, which is observed by the EU, and conforms to the obligations of the Paris Climate Change Agreement in the global context.

One of the key challenges to the transition to a climate-neutral society, environmental sustainability, and promotion of green economy is the assurance of social justice. Social justice in the globalized world and the combat against extremal inequality are a few of the essential points having a huge impact on the health and welfare of society, freedom and safety, justice and fair process, and development and realization of technologies.

The conception of inequality itself is multidimensional and requires particular attention in the modern context when analyzing "fair" possibilities of the distribution and use of energy without discrimination but involving all concerned groups in these processes. In this context, it is particularly important to find a balance among three competing goals: the access to energy services (affordability) and environmental sustainability.

Meanwhile, socio-economic vulnerability and extremal inequality, which is able to cause negative economic, ecologic, social, and psychological consequences, become one of the most relevant global issues imposing a threat to the sustainability of the whole world.

In the influence of the global pandemic situation, after stopping of industrial activities and predomination of stagnation almost in all areas of life, two opposite trends emerged: on the one hand, a significant reduction in emissions is observed; on the other hand, a huge economic recession has been experienced. A problem resulting from that is how the transition to a climate-neutral society should undergo in order to harmonize the Green Deal and ecology as well as a prosperous economy in the future and simultaneously to ensure a reduction in economic inequality.

Case of Lithuania. The themes of the interaction between inequality and sustainability in Lithuania have been poorly studied. There is a lack of studies revealing the role of inequality and its impact in the process of the transition to a climate-neutral society, assessing the real interaction between inequality and sustainability, identifying the means how to reduce or to eliminate fully the uneven distribution of environmental resources without causing any damage to the global economy and welfare of society.

The surveys conducted by the Lithuanian Department of Statistics (2019) show that one-third of the population is not able to heat their housing sufficiently because of small income. Such a situation is one of the worst in the whole Europe, while a worse one is only in Bulgaria (Eurostat, 2018).

Although in 2020, Lithuania prepared the National Energy and Climate Action Plan for 2021–2030, according to which it is intended to achieve 45% of renewable energy resources in the final energy consumption up to 2030 m., including 45% of electricity and 90% of energy in the centralized heating sector will be produced from renewable energy resources; however, an element of energy justice is ignored there.

Firstly, the expenditure on energy and energy resources in Lithuania is one of the highest in the EU community, and energy consumption in the cost of a product is 20% higher than an average in the EU.

Secondly, Lithuania is one of the EU Member States most suffering from energy poverty. In Lithuania, this indicator is 28% and exceeds even four times an average of the EU (7%) (Eurostat, 2018).

Although the data submitted highlight the content of the problems of Lithuanian energy poverty episodically only, however, it is obvious that it will not be able to ensure energy justice, consumer awareness, sustainability in all areas of life, and becoming a climate-neutral society until energy poverty has not been minimized. All this requires the conduction of further research in order to find the most suitable measures for the reduction of energy poverty and the assurance of sustainability.

It is advisable to emphasize that the impact of economic inequality on socio-economic progress is not defined unambiguously and that is confirmed by a variety of research. Most frequently, scientists study the influence of income inequality on the economic growth of the countries. Meanwhile, the studies assessing the influence of economic inequality on socio-economic progress are rarely done. Although it is often declared that the impact of economic inequality is being assessed on economic growth, however, the scoreboard of indicators highlights that the area assessed is significantly wider and includes not only economic indicators but also political, national and personal welfares, labor market, and fiscal policy indicators, which encourage the necessity to research the impact of inequality on the general socio-economic progress (and not only on economic growth).

Scientists carrying out an analysis of the impact of economic inequality on socio-economic progress (as a rule, on its individual components) apply different models reflecting fragmented areas. Besides, methods of assessment and data sources are different and the variety of data selection is restricted by their accessibility. It is stated that the impact of economic inequality on socio-economic progress may depend on a different stage of development in the states, distribution policies pursued by the state, modes of action through different channels, etc. Therefore, in order to assess the impact of economic inequality on socio-economic progress, it is required to revise (specify) the conception of economic inequality, which will comprise a methodological framework for the assessment of the impact of socio-economic progress, allow distinguishing normal and abundant inequalities, and, on this base, assess the impact of economic inequality on socio-economic progress.

1.6 Conclusions of section I

In summary, it might be noticed that the origin of economic inequality is related to the demographic changes, which took place at the end of the 18th century, subsistence needs, and poverty. It was mass poverty that influenced the changes in the political system of that time and a reduction in the popularity of the rich (nobility). At the beginning of the 19th century, economists noticed that a small social group of the rich pocketed the major part of the overall income. At that time, continuously increasing income taxation (an embryonic stage of a progressive taxation system) was offered as a solution to the problem. The imbalance of equilibrium in the market, uneven distribution of income and assets, and unrestricted accumulation of capital are the key problems that prevailed since the beginning of the 19th century. Similar issues remain in the modern world as well.

Therefore, consistent research of the phenomenon of economic inequality is particularly important in order to separate when economic inequality is justified, reasonable, i.e., normal, and when it is unjustified, unreasonable, i.e., abundant. That will be a justified set of methodologies to assess the impact of economic inequality on the socio-economic progress of the states and to find appropriate methods to reduce economic inequalities by applying them practically.

The content of economic inequality distinguishes for its broadness and multifacetedness, while the studies of economic inequality are related to the essence and content of equality and justice, equality between opportunities, and results. In order to achieve public welfare, it is particularly important to reduce economic inequality and to encourage and strengthen equal opportunities among different individuals; and, in order to ensure equal opportunities, inequality should be minimized so that people could fully realize their potential. It is significant to understand that an individual is not only a consumer of goods and services, but also a creator not only of the same goods and services but also of his life and personality. In order for a person's potential to have as many opportunities as possible for its development, appropriate conditions must be provided. Consequently, the content of economic inequality is an integral part of the country's economic situation and, simultaneously, of the welfare of society. In this context, a mutual relationship is important, i.e., the better conditions are for self-realization of people and achievement of goals, the higher potential is for the economic growth of the country, the more opportunities are provided for assurance of the quality of life of society.

Economic inequality cannot be explained on the basis of one specific economic theory just like the reasons for the deepening of economic inequality cannot be summarized in one specific factor or specific circumstances. Economic inequality is determined by a complex matter of factors, such as economic, social, political, demographic, health, and psychological factors. Thus, the roots of the issue are not in only one fragmented factor but depend on a series of factors. The fight against inequality requires a systematic approach, a complex solution in most areas starting with reforms of financial and anti-monopoly policies, and should finish with legal regulation of education and employment relationship.

Abundant inequality is one of the acute and dramatic themes in modern society. The concepts of normal and abundant inequalities are only being formed lately. The issues related to inequality are analyzed by investigating income distribution, meanwhile, there are very few studies distinguishing a differentiated point of view toward economic inequality itself, its impact on the assessment of countries' economy through a prism of normal and abundant inequalities. The latter trend of research is relatively new, and, first of all, it is quite difficult to distinguish normal inequality from abundant one; therefore, there are only a few methods of assessment, and they are not widespread and well-known.

The authors of this monograph offer a differentiated approach toward economic inequality. A certain degree of inequality can be justified if improvement, competition, and investment in the welfare of society are encouraged. Meanwhile, inequality becomes unjustified when its certain level does not encourage anymore but contrarily restricts economic growth and calls out negative socio-economic consequences. So, inequality can be normal and abundant. Abundant inequality becomes a feature of economic and social distortions restricting the economic growth of the states and negatively affecting the quality of the population's life.

Economic inequality is a serious problem of the whole world, and the extent of this problem is mostly reflected by wealth inequality. The gap between the richest and the rest of society continues widening as the richest planet's population receive the major benefit from the growth of economics, and the assets accumulated by them is rapidly increasing. Global economists emphasize that inequality is the result of economic policies pursued. The concentration of tangible assets in the hands of one group not only destroys the principles of social justice, but it also restricts the opportunities of the members of the rest of society and the quality of their life, and simultaneously suppresses and discourages the socio-economic progress of the state itself.

It is worth highlighting that the impact of economic inequality on socio-economic progress is not defined and assessed unambiguously. This is confirmed by a variety of research. The influence of income inequality on the economic growth of the countries is investigated most frequently. Meanwhile, scientific studies assessing the impact of economic inequality on socio-economic progress are rarely met, and the indicators applied in empiric stories should be noted as well. Although it is often declared that the impact of economic inequality is being assessed on economic growth, however, the scoreboard of indicators highlights that the area assessed is significantly wider and includes not only economic indicators, but also political, national and personal welfares, labor market (occupation), and fiscal policy indicators, which encourages to research the impact of inequality on the general socio-economic progress, as a whole and not only on economic growth.

Scientists carrying out an analysis of the impact of economic inequality on socio-economic progress use different models, apply different methods for assessment, and justify the analysis on the basis of different variables. Moreover, the periods selected for research and data sources differ and the variety of data selection is restricted by their accessibility. It is stated that the impact of economic inequality on socio-economic progress may depend on a different stage of development in the states, distribution policies pursued by the state, modes of action (channels), etc. Therefore, in order to assess the impact of economic inequality on socio-economic progress, it is required to revise (specify) the conception of economic inequality which will comprise a methodological framework for the assessment of the impact of socio-economic progress, allow distinguishing normal inequality from abundant, and assess the impact of economic inequality on socio-economic progress.

2 METHODOLOGY FOR ASSESSING EXCESSIVE INEQUALITY

2.1 The concept of economic inequality based on the relationship between normal and excessive inequality

Inequality in the distribution of resources has traditionally remained one of the most pressing challenges in the economy. However, in recent decades (especially after the global crisis in 2008), special attention of world economists, Lithuanian scientists, and politicians, has been paid to economic inequality. The current high level of income differentiation is recognized as a negative factor with negative socio-economic consequences. The imperfection of redistribution levers created conditions for excessive concentration of resources and material opportunities in a small group of the population.

When it comes to inequality, first of all, it means that the economy and society of the country are rich and poor. Moreover, in classifying people, societies, or countries as rich, the emphasis is not so much on income as on wealth. Revenue shows how much a person's purchasing power has increased over a period of time, and assets determine the volume of assets over a given period.

Economic inequality is differentiation according to which individuals, social groups, and classes are in different vertical socio-economic hierarchies (stages and degrees) and have different life chances and needs to meet, different access to public goods and services (education and science, health, culture, leisure, social security, etc.). Generally speaking, economic inequality is a differentiation of quality of life.

Inequality is a form of economic differentiation. Therefore, economic inequality is the difference in income and living standards between individuals and groups.

Economic inequality is a difference that arises in various ways of measuring economic well-being between individuals, groups, societies, and countries. Sometimes economic inequality is perceived as income inequality, wealth inequality, or wealth gap. Economists typically focus on inequalities in three dimensions: wealth, income, and consumption. Research projects on economic inequality are related to the essence and content, equality of results, equality of opportunities.

DOI: 10.4324/b22984-3

Socio-economic inequality remains a concept of breadth and complexity. The concept of inequality is very simple and at the same time very complex (Salverda, Nolan, & Smeeding, 2013). It is simple and even because in different times of history it has forced different people from different social backgrounds to migrate around the world. But, at the same time, it is very difficult because it has forced even the most talented philosophers, political theorists, sociologists, and economists to find out its meaning and possible implications. Economic inequality can be understood as inequality affecting the economy, or inequality of economic origin, which can be both a consequence and a cause of existing economic processes (Salverda, Nolan, & Smeeding, 2013).

In view of the above-mentioned forms of inequality, it can be said that all of them are an integral part of the economy, which can hinder or promote economic growth and have certain consequences.

Income differentiation – the key economic variable – is one of the main problems of economic inequality and a major focus of most inequality-related research, while low income is undoubtedly one of the key indicators of poverty levels. However, income is only one component of this problem. The amount of accumulated assets is also a part of economic resources. In terms of income and wealth distribution, poverty can lead not only to exclusion of people in terms of income and wealth, but also to exclusion in social life (Salverda, Nolan, & Smeeding, 2013).

The assessment of income inequality is inseparable from the concept of social justice. It is an agreement within society in a country, which is right for us. Also, social justice is not defined differently from individual theoretical perspectives, although all theories of justice are based on the premise that people are equal and try to explain the nature of that equality. For example, formal justice is understood as the equality of all before the law. According to utilitarian theory, everyone is different and therefore there are different achievements. Proponents of a meritocratic approach support a utilitarian approach and argue that the focus is on merit. The compensation theory, meanwhile, argues that people should be compensated for the harm done by their origin and nature and that equality between people should be increased (Lazutka, 2003).

The concept of inequality itself, together with the concepts of poverty or exclusion, can be interpreted in different ways. The literature has repeatedly tried to distinguish the differences and features of these concepts, but there was no consensus. On the other hand, it has led to improved tools for measuring and measuring inequality. Atkinson's work *On the Measurement of Inequality* (1970) and A. Sen's scientific paper on poverty measurement tools opened up opportunities to present the latter phenomenon in further literature and to fundamentally change empirical research on this problem (Salverda, Nolan, & Smeeding, 2013).

The empirical analysis focused on an accurate assessment of the extent and trends of inequality, especially in terms of earnings and income. As highlighted by Jenkins and Micklewright (2007), the development and refinement of the nature and quality of available data have become the basis for these studies. Easily accessible micro-level data, in contrast to grouped data from public sources, has become crucial.

Considerable efforts have also been made to harmonize inequality assessment tools and concepts for intercultural use, as it was in the case of the *Luxembourg Wealth Study (LWS)* in Luxembourg when the organization's databases included data from OECD countries and Eurostat statistics. The increasing amount and long-term availability of long-term data has also provided an opportunity for inequality research to gain a foothold in recent literature.

The literature pays a lot of attention to the explanation of the main causes of economic inequality and its persistence. However, the relationship between theoretical and empirical inequality research material is often weak compared to other areas of economics, as the practical part of research favors careful processing of existing data. It should be noted that research on economic inequalities can sometimes appear to be far removed from the general field of economic research, especially for economists looking at other areas. This explains Atkinson's (1997) attempt to achieve universal recognition of research on income inequality.

On the other hand, economists can draw a wealth of experience from the link between this problem and other sciences, such as sociology, which areas of research include inequality, its causes, and consequences. It is to be hoped that current trends in socio-economic research, such as the juxtaposition of households and the labor market, the promotion of women's participation, and the reduction of unemployment, as well as increasing the link between income and employment globally, will allow inequality research to become more entrenched in the economic discipline.

Inequality – is the conditions under which people have unequal access to social goods – money, will, and prestige (Smeltzes). Inequality exists in all societies, even in the most primitive ones. In more advanced societies, inequality becomes more pronounced. The essence and causes of inequality are treated differently. The causes of inequality can be:

- *Different income levels of social groups.* Income differentiation is due to a variety of factors: first, linked to personal abilities and achievement; second, economic, demographic, social, health, and psychological health and political factors can act independently of them.
- *Uneven degree of adaptation of different social groups to new living conditions.* This is typical of both privileged groups that have had unlimited access to a variety of goods and services in the past, and non-elite social groups accustomed to paternalistic state care.
- *Regional and territorial characteristics that influence the solution of socio-economic issues.* Regional and intersectoral wage differentiation is also a factor in inequality.

- *Unequal distribution of property and efficiency of its use.* "Wealth attracts wealth" – this thesis is characteristic of the differentiation of wealth and income. Disposal of a property provides an opportunity to receive income from property – dividends, etc.
- *Demographic factors.* The size and composition of the family, age, and place of residence can strongly determine the level of income and material conditions of various families and individuals.
- *Gender.* It is well known that the distribution of income by gender is different. The pay gap between men and women is affected by the intersectoral differentiation of economic activities – there are male and female occupations. In general, wages in men's work areas are higher than in women's.

EU policy aims to ensure a high quality of life, a standard of living in line with human dignity, and to create opportunities for living in an active, integrated and healthy society. *Only social objectives, public welfare, and quality of life can be expected to boost the country's economic growth, modernize business, increase competitiveness,* and, at the same time, create a sustainable source of public revenue and tackle recent challenges such as unemployment, regional disparities and economic inequality in the EU (Rakauskiene et al., 2015).

After analyzing contemporary scholarly approaches to inequality, the author of this monograph presents *a concept of economic inequality that consists of key provisions and economic conceptions of economic inequality.*

The main *provisions* of the concept of economic inequality proposed by the authors are the following:

- Economic inequality is the distribution of income, consumption, savings, material living conditions, wealth, and different opportunities for access to public goods (education, health, services, recreation, culture, and social services) depending on economic, social, demographic, psychological factors, and abilities on macro (state) and micro (social groups and individuals) levels. Economic inequality is not only the result of different social, demographic, and economic factors but also the result of economic policies. "Inequality is not inevitable, it is the result of political decisions" (J. E. Stiglitz). "Wealth and income inequality are not only the result of the economy but also of politics" (T. Piketty).
- Economic inequality can be normal (reasonable) and excessive (unreasonable). A certain degree of inequality is justified, depending on education, qualifications, responsibility, as well as on the incentives created for innovation and entrepreneurship, on development, competition, and investment promotion. However, growing economic inequality becomes a problem when it restricts individuals' educational or professional choices, and reduces opportunities for self-realization when individuals' efforts are directed only at meeting the most basic needs. Excessive

inequality is not just deep inequality, but one that, from a certain level of achievement, begins to hamper socio-economic progress.

- Economic inequality traditionally includes the differentiation of income and consumption, but also differences in material conditions and distribution of wealth in society. The scientific literature usually distinguishes between two concepts – income inequality and wealth inequality. These concepts are interrelated, as often the level of assets available and the ability to acquire them depend directly on income. Although the authors do not consider this division to be exhaustive, and given that both income inequality and wealth inequality are components of economic inequality, there are other, no less important components of this phenomenon: consumption, savings, and access to public goods and services. The main goal of this study is to distinguish normal from inequality from excess, using only income differentiation data. The unequal distribution of income in the current generation leads to unequal opportunities for future generations as well. Children of poor parents have much smaller and lower-quality opportunities for access to education and science, skills development, access to culture, and health services, which may lead to future and limited conditions of their professional careers, lower material, and social status.
- Socio-economic progress includes development, emphasizing the synergies between the three dimensions of economic growth, quality of life, and sustainability. Therefore, when assessing the impact of economic inequality on socio-economic progress (EU-28), it is appropriate to analyze the interaction of economic growth, quality of life, and sustainability factors with indicators of economic inequality.
- After identifying the criteria for assessing economic inequality and socio-economic progress, forming preconditions for separating normal inequality from surplus, it is possible to assess the impact of economic inequality on socio-economic progress – economic growth, quality of life, and sustainability.
- The proposed concept is an econometric model, the basis for distinguishing between normal and excessive inequality, and for assessing its impact on socio-economic progress.

The concept and structural basis of the socio-economic inequality proposed by the authors (Figure 2.1) consists of *three levels.*

The first level – is the factors that cause economic inequality. The second level highlights a differentiated approach to economic inequality, emphasizing the relationship between normal and excessive inequality. The third level – is the assessment of the impact of economic inequality on the socio-economic progress of EU countries (see Figure 2.1).

The first structural level of the concept of economic inequality – is the social, economic, and environmental factors that influence economic inequality.

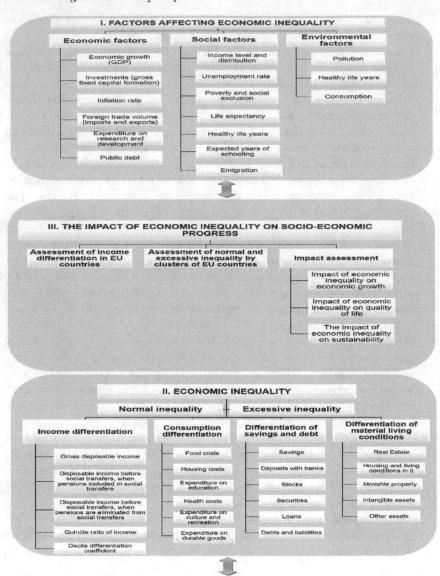

Figure 2.1 The concept of economic inequality based on the relationship between normal and excessive inequality.

Source: Compiled by the author.

Economic inequality is influenced by economic, social, and environmental factors:

• Economic factors include economic growth (GDP) and investment (gross fixed capital formation), inflation rate, foreign trade volume, research expenditure, and public debt.

- Social factors include a level of income and distribution of income, unemployment rate, poverty, and social exclusion, life expectancy, healthy life years, average education, emigration, etc.
- Environmental factors include environmental pollution, government spending on the environment, consumption, forest area, and consumption of fossil fuel resources.

The content and structure of economic inequality are revealed by the typology presented at the second level. It consists of the differentiation of income, consumption, savings and debts, and material living conditions, assessed according to selected indicators.

One of the most important indicators of inequality is a differentiation of income. It occupies a special place in economic inequality. These are real differences in the income of the population, the distribution of the population according to the size of income. Income differentiation also largely determines economic inequality. Monetary income mainly determines the well-being of the population, work motivation, and the socio–economic situation of people and the political situation in society depends on them.

A society with a rational, normal income differentiation is the most stable because it consists of a large middle class, and it is characterized by high social mobility, strong motivation to pursue a professional career, and high social status. Conversely, societies with high, polar income differentiation are socially and economically unstable, and the promotion of professional growth is low, with high levels of crime.

Therefore, economic inequality is the difference in material conditions between people and social groups, and the ability to meet their own needs, based on income differentiation.

Income is a very important, but not the only indicator of high quality of life. The well-being of households depends on the number of family members, their health, the economic and social environment, and economic resources. The Gini distribution of disposable income in OECD countries ranges from 0.30 to 0.50, while the overall Gini distribution of assets ranges from 0.50 to 0.80 (Davies, 2013).

The closest equivalent to income is consumption expenditure. Consumption is a more accurate indicator of well-being than income simply because well-being generally reflects the direct consumption of goods and services. On the other hand, it is emphasized that it is extremely difficult to calculate the actual level of consumption in rich countries. Consumption expenditure data are usually collected to calculate the consumer price index rather than the volume and volume of consumption. Only in rare cases, the aim is to find out the true amount of consumption, considering that durable goods such as household appliances, cars, and housing are bought by people during one period of life but often used by another. Indeed, for certain groups of people, such as older people living in their own homes and not paying their home loans, the real consumption expenditure may differ significantly from the consumption expenditure estimates. It can be argued, firstly, that it is appropriate to prioritize research overconsumption rather than consumption, and secondly, rather than a one-off (short-term) rather than permanent income.

The dominance of "strategic" expenditures in social structures aimed at strengthening human capital (investment in education, acquisition of real estate, and savings) is characteristic of affluent social groups. Meanwhile, consumption expenditures of vulnerable social groups are directed toward meeting the current necessities in order to "survive".

A good supply of real estate can be the basis of prosperity in affluent social groups. For example, housing and real estate create the conditions for *the accumulation of savings*, which can be passed on to the next generation or converted into financial capital. In poor families renting a house, housing is not active; therefore, the costs of housing maintenance are characterized by cost features.

One of the most important aspects is the differentiation of material living conditions and wealth. The topic of wealth inequality is relatively new and has not been fully explored due to data availability and measurement complexity. Asset differentiation has recently been especially emphasized by the world and Lithuanian experts. However, the difficulties in valuing assets are illustrated by the following reasons. Firstly, the valuation of assets involves a large amount of data from a variety of sources, which is very difficult to compare. Secondly, the problem of asset valuation is related to the confidentiality and availability of data (such as the transaction amounts of assets acquired or sold are not available, and the tightening of personal data protection has further complicated asset concentration assessments). Therefore, the assessment of asset differentiation causes a huge problem in the economy.

Accumulated and newly created assets are the basis for increasing the country's economic potential, determining its development opportunities. The concept, content, and forms of property, as well as property, have changed over time depending on economic relations and social realities. Land, as an object of property, arose when people became sedentary and began to engage in agriculture. With the advent of new industries and industries, new objects of property rights have emerged. With the transition of society from an agricultural-based economy to the industrial era, intangible objects have also become objects of ownership, i.e., scientific inventions, literary, and artistic works.

Finally, as society transitions from an industrial to a knowledge-based economy, new forms of assets emerge, in addition to traditional intellectual property, such as intellectual capital, which consists of all the intangible assets of an entity, whether or not they can be accounted for and regulated. Intellectual capital, consisting of information and knowledge, is becoming the most important asset of the knowledge economy in today's world.

In modern society, both human resources and material assets play a particularly important role. However, financial capital is simply necessary to start your own business in order to get a high return on investment, and a certain amount of savings is relevant in case of job loss, divorce, or unforeseen medical expenses.

Assets are also important because they give the holder more freedom and empowerment. It is very important for an individual seeking to defend their rights to be able to hire the best lawyers. Funding political campaigns, lobbying, and influencing politicians to defend certain interests – all of this also requires significant funding. On the other hand, great wealth gives the power to do good things. The world's richest people, from Andrew Carnegie to Bill Gates, who have achieved impressive success, have dedicated much of their vast wealth to philanthropy (Davies, 2013).

The term "asset" is used to mean all tangible assets (excluding arrears). Theoretically, the concept of assets includes cash, bank deposits, other liquid assets, stocks and bonds, business shares, owner–occupied real estate or any other real estate, as well as durable goods, including antiques, art, and jewelry. In practice, however, not all forms of property are included in this definition; e.g., the concept of durable goods is often limited to cars. In addition, the concept of property itself, as well as its forms such as durable goods and real estate, are often understood differently in different countries.

Wealth inequality largely depends on income alone. This factor also affects the consumption ratio. Since the 1960s, when the first surveys on the distribution of income and wealth were conducted, it became clear that income and wealth were two interrelated but not completely identical concepts. Their correlation coefficient is approximately 0.5 (Davies, 2013). One of the factors that make the difference between these two concepts is age. If we compare the two concepts in terms of age graphically, you would notice that the income curve would rise much earlier than wealth. People of retirement age receive relatively little income but have a lot of wealth, and conversely, young people may have a high income but no wealth. In addition, the self-employed may have high equity but low income (albeit temporarily).

However, *the main focus of the second level of the concept of economic inequality is the relationship between normal and excessive inequality.* Bearing in mind the above arguments in this work, in order to achieve the goal of how to estimate normal inequality from surplus, the authors limit themselves to income differentiation data. The methodology develops the preconditions for identifying normal and excessive inequalities, finds breaking points and identifies a marginal effect to observe when economic inequalities turn from normal to surplus and thus negatively affect the socio-economic progress (economic growth, quality of life, and sustainability) of the EU-28. In distinguishing normal inequality from surplus, the empirical study uses the following indicators of income differentiation: total disposable income, disposable income before social transfers when pensions are included in social transfers (it means income of population excluding pensions and other social benefits), disposable income before social transfers when pensions are eliminated from social transfers (income of the population together with pensions, but without other social benefits), and general indicators of income differentiation (decile differentiation coefficient and income quintile ratio).

The third structural level of the concept of economic inequality is the assessment of the impact of economic inequality on socio-economic progress in the EU-28. After analyzing the state and development of income differentiation in the EU countries, groups of countries with normal and excessive inequality, which form certain clusters, were distinguished. Establishing the most appropriate criteria for assessing economic inequality and socio-economic progress provides an opportunity to assess the impact of economic inequality on socio-economic progress – growth, quality of life, and sustainability – in the EU-28 and to quantify the break points when economic inequality arises.

2.2 Justification of excessive inequality methodological assessment according to EU clusters

It should be noted that quantified normal and excessive concepts of economic inequality are still emerging. The authors propose three methods for calculating excessive inequality:

I method. It is assumed that economic inequality with a decile differentiation coefficient (K_d) of no more than seven times is justified (treated as normal) in European countries. By calculating the decile differentiation coefficient (K_d), it is possible to calculate the excess inequality and divide the EU-28 into three clusters:

- countries with high excess inequality with $K_d \geq 10$;
- countries with low and medium inequality when (K_d 10 $<\geq$ 7); and
- countries of normal inequality with $K_d <7$.

II method. The concept of economic inequality is closely linked to the concept of poverty. After finding the optimal poverty line ($Sk_{optimal}$) and knowing the real poverty statistics (Sk_{real}), the Excessive Inequality (PN_{Sk}) is calculated according to the formula below, assuming that the unjustified (excessive) inequality is the difference between the existing (actual) optimal and the tolerable level of poverty. The EU-28 is divided into three clusters:

$$PN_{Sk} = Sk_{real} - Sk_{optimal}$$

- countries with high excessive inequality ($PN_{Sk} > 7$);
- countries with small and medium inequality (PN_{Sk} 7 $\leq \geq$ 4); and
- countries of normal inequality ($PN_{Sk} <4$).

III method. Unjustified (excessive) economic inequality is interpreted as exceeding the existing part of the inequality as a justifiable (normal) part of the inequality. In this context, the Gini coefficient is the most appropriate for the analysis of economic inequality. After finding the optimal inequality threshold ($Gini_{optimal}$) and knowing the real inequality statistics ($Gini_{real}$), the excess inequality (PN_{Gini}) is calculated according to the formula below, assuming that the unjustified (excess) inequality is the difference between the existing (real) tolerance and the real Gini coefficient level. The EU-28 is also divided into three clusters:

$$PN_{Gini} = Gini_{real} - Gini_{optimal}$$

- countries with high excessive inequality ($PN_{Gini} > 4$);
- countries with low and medium inequality (PN_{Gini} $4 \leq \geq 0.1$); and
- countries with a normal level of inequality ($PN_{Gini} < 0$).

The chapter assumes that indicators of the economic and quality of life in the Scandinavian countries should be considered as corresponding to the optimal/normal level. The Scandinavian countries were chosen as the target for the following circumstances:

1 Geographically, the Scandinavian countries and Lithuania are not very far away, which leads to cultural and mental similarities.
2 Scandinavian countries have high quality of life standards.
3 The Scandinavian countries have a high quality of life among the EU countries. The Scandinavian openness of the countries, the tendency to free trade, and the implementation of free trade policies have made a significant contribution to economic success.
4 In Scandinavian countries, social security policies are aimed at supporting the middle class. There is a strong focus on social benefits, which are targeted at the poor class.
5 The priority areas for the Scandinavian countries are quality education and health care. In addition, corruption was managed and a transparent governance provision was implemented. This has created an opportunity for the effective development of welfare programs.
6 According to global surveys, the Scandinavian countries are a home to the happiest people in the world. The happiness index of the Finnish population was 7.86, that of Denmark was 7.65, and that of Sweden was 7.37 (the happiness index of the Lithuanian population was 6.31 in 2018, and the EU-28 average was 6.62).
7 The Scandinavian countries have a proactive public stance to reconcile issues of economic efficiency and social justice. The prevailing Scandinavian welfare model is based on high social capital, social trust, and a level of social solidarity. Nordic countries lead the world not only in such social indicators as social equality, social development, and social reintegration but also in economic indicators – countries' competitiveness, level of a knowledge economy, and innovation.

In view of all the circumstances above, the experience of the Scandinavian countries is considered as a role model.

In this way, the assessment of normal (justified) and excessive (unjustified) inequalities is carried out using the three different methods proposed in order to compare the results obtained and to determine in which EU countries the excessive inequalities are most prevalent. The identified indicators of the level of excessive inequality are also used in the impact assessment of the economic inequality on socio-economic progress.

2.3 Difficulties in assessing wealth inequality and proposed methods

Differentiation of income – is the economic variable – one of the most important problems of socio-economic inequality and the main object of most research related to inequality. However, cash flow is only one component of this problem. *Material living conditions, the accumulated amount of wealth, and its distribution are much more eloquent indicators.* In terms of the unequal distribution of income and wealth, poverty can lead not only to exclusion of people in terms of income and wealth, but also to exclusion in social life (Salverda, Nolan, & Smeeding, 2013), thus preventing life satisfaction and quality of life.

Material living conditions and their distribution in society can be analyzed using a variety of variables. The main variables are various financial indicators, such as expenses, income, and assets, as well as non-financial indicators, such as various indicators of material well-being, happiness, and satisfaction.

It should be emphasized that assets are more unevenly distributed compared to human resources, profits or income. The distribution of disposable income according to the Gini coefficient in OECD countries ranges from 0.30 to 0.50, while the overall distribution of wealth according to the Gini coefficient ranges from 0.50 to 0.80 (Davies, 2013). Currently, 1% of the richest US people own more than 30% of the country's assets, as well as, according to estimates, 2% of adults own 50% of the wealth of all households in the world. Wealth and income correlate with each other, and the aforementioned high level of concentration highlights the problem of socio-economic inequality (Davies, 2013).

In modern society, both human resources and material assets play a particularly important role. However, financial capital is simply necessary to start your own business in order to get a high return on investment, and a certain amount of savings is relevant in the event of job loss, divorce, or unforeseen medical expenses. Property is also important in that it gives its holder more freedom and empowerment. It is very important for an individual seeking to defend their rights to be able to hire the best lawyers.

Table 2.1 shows the results of two international surveys of wealth inequality.

Table 2.1 International comparison of wealth inequality

Country	UNU-WIDER Global Wealth Study (Davies et al., 2007)				The Luxembourg Wealth Study (Sierminska et al., 2006)			
	Year	Share of top 10%	Share of top 1%	Gini	Year	Share of top 10%	Share of top 1%	Gini
Austria	2002	45.0		0.622				
Canada	1999	53.0		0.688	1999	53.0	15.0	0.750
Denmark	1975	76.4	28.8	0.808				
Finland	1998	42.3		0.621	1998	45.0	13.0	0.680
France	1994	61.0	21.3	0.730				
Germany	1998	44.4		0.667	2002	54.0	14.0	0.780
Ireland	1987	42.3	10.4	0.581				
Italy	2000	48.5	17.2	0.609	2002	42.0	11.0	0.610
Japan	1999	39.3		0.547				
Korea	1988	43.1	14	0.579				
New Zealand	2001	51.7		0.651				
Norway	2000	50.5		0.633				
Spain	2002	41.9	18.3	0.570				
Sweden	2002	58.6		0.742	2002	58.0	18.0	0.890
Switzerland	1997	71.3	34.8	0.803				
United Kingdom	2000	56.0	23	0.697	2000	45.0	10.0	0.660
USA	2001	69.8	32.7	0.801	2001	71.0	33.0	0.840

Source: Davies (2013).

Based on these data, it is clear that there are significant wealth inequalities in all 17 OECD countries. Ten percent of the population depends on 39% up to 76% of total assets, and the Gini coefficient ranges from 0.55 to 0.81. For comparison, Davies et al. (2007) in the study of the distribution of disposable income, the Gini coefficient ranges from 0.30 to 0.50 in all 17 countries (Davies, 2013). It is worth mentioning that the results presented in Table 2.1 were obtained using different methods: surveys, property tax data, and estate multiplier method. Given the variety of methods used, it is not clear what effect intercultural differences had on the results of the first column of Table 2.1. The highest scores are likely to have been obtained in the United States, where it turned out to be 10% of the population gets 69.8% of the total country's assets, and the Gini coefficient is 0.801. In several other countries where the survey method was used, the results were similar. Based on research data from property tax and inheritance valuation charts, it turned out that 10% of the UK population own 56% of the property, 10% of the Swiss population own 71.3% of the country's assets, and the Gini coefficient ranges from 0.697 (the United Kingdom) to 0.803 (Switzerland). It is worth noting that data obtained from fundamentally completely different research methods are not the only known international data. OECD countries are currently working closely together to develop a common approach for all countries to compare and analyze wealth data internationally (*Luxembourg Wealth Study*).

In the second column of Table 2.1, the preliminary results of the above-mentioned study are presented. All the results were obtained during surveys, the statistical unit was household in all cases, and the perception of property coincided in all cases. Unfortunately, in order to use a common concept of assets, some data obtained in almost all countries have been eliminated (e.g., data on pension accumulation accounts). It should be emphasized that both graphs in the Table 2.1 are closely related to each other, despite the different methods used. A comparison of the two counts shows that Italy has the lowest wealth inequality and the United States the highest. The countries' assessment of Gini coefficients in both the United Nations University World Institute for Development Economics Research (UNU-WIDER) and the Luxembourg Wealth Study overlap, with the exception of the United Kingdom; according to the data obtained by the latter, it has a lower Gini coefficient. This difference was due to different survey methods: UNU-WIDER used the property multiplier method, while the Luxembourg property study was based on data from the *British Household Panel Study* (BHPS).

Based on research from UNU-WIDER and the Luxembourg Wealth Study, countries can be divided into the following categories according to the distribution of wealth and the Gini coefficient (Davies, 2013):

1 countries with high wealth inequality: Sweden and the United States;
2 countries with medium-level property inequality: Canada, Germany, and the United Kingdom;
3 countries with low wealth inequality: Italy and Finland.

Oddly enough, Sweden falls into the same category as the United States. There is a simple explanation for this – despite the fact that 1% and 10% of the richest people in the country account for a smaller share of the country's wealth than in the United States. A very large number of people in Sweden still have little or no wealth (according to official statistics, as many as 30% of the country's population is poor). The accuracy of these results is difficult to verify, but they have undoubtedly been influenced by the high level of indebtedness, the low percentage of property owners, and the supposedly highly developed pension accumulation system. For example, Finland, like Sweden, is in the last category with high pensions and social welfare. This may have been due to a variety of reasons, but the main ones were a lower net present value of cash flows and a higher percentage of property owners (Davies, 2013).

The impact of wealth inequality on economic well-being depends on that wealth and income. This factor also affects the asset-consumption ratio. Since the first surveys in the 1960s on the distribution of income and wealth were conducted, it became clear that *income and wealth were two interrelated but not completely identical concepts. Their correlation coefficient is approximately 0.5* (Davies, 2013). One of the factors that makes the difference between these two concepts is age. If we compare the two concepts in terms of age graphically, you would notice that the income curve would rise much earlier than wealth. People of retirement age receive relatively little income but have a lot of wealth, and conversely, young people may have a high income but no wealth. In addition, the self-employed may have high equity but low income (albeit temporarily).

The forms of expression of income and assets are similar in most cases. For example, financial value is one of the main forms of expression of both assets and income. However, the percentages of equity and homeowners differ significantly. According to US consumer surveys, in 2004 about 40.3% of the country's first-quantile families had their own housing, and only 15.2% of people belonging to the first quartile group had their own housing. In the case of equity, the contrast is even more obvious: 45.8% of the richest people in the first decile have their own businesses, while, according to income distribution data, only 34.7% of people in the first decile have their own businesses. These results confirm that business-related taxes have some effect on such a distribution (Davies, 2013).

It should be noted that a number of significant empirical studies have already been conducted to investigate the impact of income and wealth on consumption. An empirical model has been developed to substantiate the equitable consumption hypothesis (Davies, 2013). Income-related economic changes have a direct impact on consumption, and various ways to increase liquidity, such as lottery winnings, should have a positive effect. However, in practice this is not always the case, as assets are an endogenous variable: the lower the time priority, the more assets it has. In order to determine the real impact of assets on consumption, it is necessary to find a suitable instrumental variable for valuing assets. Rising stock and real estate prices had a rather positive effect on consumption. Rising stock and real estate prices had a fairly positive effect on consumption (Davies, 2013). This has raised concerns that a market downturn will lead to lower consumer spending and increase the likelihood of a recession, which is what happened.

Property evaluation methods and issues. Foreign researchers use three subdimensions to assess income and economic exclusion: (1) material deprivation; (2) food security, housing, and financial risk or economic hardships; and (3) financial well-being.

Data on material living conditions of the Lithuanian population have been collected by the Lithuanian Department of Statistics since 2005, implementing Regulation (EC) of June 16th No 1/2003 of the European Parliament and of the Council 1177/2003 concerning Community statistics on income and living conditions. The survey collects data on household income, employment, housing conditions and problems, assessment of living conditions, and the ability to meet certain needs. The Department of Lithuanian Statistics has also published the indicators of Lithuanian life quality. The ones describing material living conditions are the following:

1   *The level of material deprivation (percent)* – an indicator showing the share of the total population who face certain elements of material deprivation due to lack of funds. *A person is considered to be in severe material deprivation if he or she encounters at least four of the nine elements of material deprivation.* The list of items of material deprivation includes five economic difficulties and four durable items that the household does not have due to lack of funds.

Economic difficulties:

1   The household is unable to pay housing rent, utility bills, housing or other loans, credit payments on time due to lack of money.
2   The household does not have the opportunity to spend at least a week away from home.
3   The household cannot afford to heat the dwelling sufficiently.
4   The household cannot afford to eat meat, fish, or similar vegetarian food at least every other day.
5   The household would not be able to pay for unforeseen expenses (the amount of expenses equal to the monthly poverty risk threshold of the previous year) from its own funds.

Long-term use items:

1   telephone, including mobile,
2   color TV
3   washing machine, and
4   car.

2   *The housing deprivation rate (percent)* is defined as the share of people living in overcrowded housing and experiencing at least one of the housing problems. A dwelling is considered to be overcrowded if the household occupies fewer rooms than the set minimum number of rooms. The minimum number of rooms is set as follows:

- one room per household;
- one room for each married couple;
- one room for each member of the household over the age of 18 not in the previous category;
- one room for two people aged 12–17 of the same sex;
- one room for each person aged 12–17 not in the previous category; and
- one room for two children up to 12 years old.

Housing problems – dripping roof, rotten windows, floors or wet walls, housing without bath or shower, toilet with running water, too dark (not enough daylight entering through windows), living space is characterized by noise, air and environmental pollution, and high crime.

The economic literature and statistical surveys also use *the concept of Intensity of Material Deprivation*, which is defined as an indicator expressed as the average number of elements of material deprivation per person who faces a minimum or higher number of elements of material deprivation due to lack of funds.

In the scientific literature, the term "property" is understood differently in legal, economic, physical, social, or other terms. Each field of science, in defining wealth, emphasizes the characteristics and priorities specific to that field. This explains why the definition of "property" in the most general sense is very abstract (Rakauskienė, 2010). In the Law on the Fundamentals of Property and Business Valuation of the Republic of Lithuania, the property is defined as tangible, intangible, and financial values. The same law distinguishes between two types of property:

1 "immovable property" means land and associated facilities whose location cannot be changed without changing their intended use or reducing their value and economic purpose, or property which is recognized as such by law; and
2 "movable property" means property which can be transferred from one place to another without altering its contents, without substantially reducing its value, or without seriously damaging its purpose, unless the law provides otherwise.

The Law on Personal Income Tax of the Republic of Lithuania defines assets as movable and immovable property, securities, and derivatives, and other intangible assets. According to the latter definition, it is clear that, firstly, a property is not only objects used to meet human needs, but also certain legal and financial instruments which express the rights of their owner and create an appropriate mechanism to require others to act in their favor. Secondly, it may be tangible or intangible. According to B. Galiniene, when structuring the concept of property, the property consists of tangible and intangible objects, and, by nature, it can be real and movable.

In modern civil law, property is generally understood as the totality of the property units belonging to a particular person; this set of property units has a specific owner in a specific period and can be individualized by that owner and separated from other property units. It should be noted that the Civil Code of the Republic of Lithuania does not provide a definition of "property"; only Article 1.97 provides that the objects of civil rights are things, money and securities, other property and property rights, results of intellectual activity, information, actions and results of actions, as well as other property and non-property values. The definition of tangible property is given in Article 4.1 of the Civil Code of the Republic of Lithuania: objects are considered to be objects of the material world appropriated from nature or created in the production process.

Meanwhile, the prevailing opinion in economic theory explains: *anything that has value and belongs to somebody should be considered as a property*. Analyzing the concept of "assets", it can be said that in each case, a larger and smaller number of individual assets or types of assets is listed, but it is not provided that the concept, types, and forms of assets are summarized. In the economic context, the following criteria of the concept of property can be distinguished (Rakauskiene, 2010):

1   the property must have a value, and
2   the entity must have the right to the object,
3   the object must provide benefits in the future.

Each *value is an asset only when it has an owner. Possession, disposal, or management of property determines a person's security (not only material but also moral), self-confidence, self-esteem, as well as the corresponding quality of life*. In this context, it is important to note that it has historically been the case that real estate should be treated as the most valued asset. It may seem ironic, but the folk exception teaches that everyone must "build a house, raise a son, plant a tree". Housing is a particularly important part of human well-being and at the same time a guarantor of social stability. However, the scientific literature still pays insufficient attention to this phenomenon – no attempt has been made to assess property differentiation, nor has it been studied how property differentiation, with an emphasis on housing, affects a person's quality of life.

Individual assets can be three main forms:

1   physical property – land, house or apartment, car, household appliances, furniture, art treasures, jewelry, etc.;
2   financial assets – shares, obligations, bank deposits, money, checks, bills, etc.;
3   human capital – talent, genius, abilities, education, mental and physical health, non-standard thinking, creativity, etc.

Figure 2.2 shows an economic classification of wealth created by the authors. In this case, material wealth is the most important factor, with the distribution and correlation between real estate and quality of life emphasizing the importance of dwelling place. Meanwhile, movable and other type of wealth is considered to be directly related to obtainable income. The opportunity to attain and manage non-material wealth, such as copyrights and licenses, in turn depends on the amount of material wealth a person has. This opportunity might emerge if a person has a certain social status or has accumulated a certain amount of material wealth, or in the case of favorable circumstances (e.g., inheritance).

Methodology for assessing property inequality of the Lithuanian population. In the analysis of economic inequality, a representative questionnaire survey

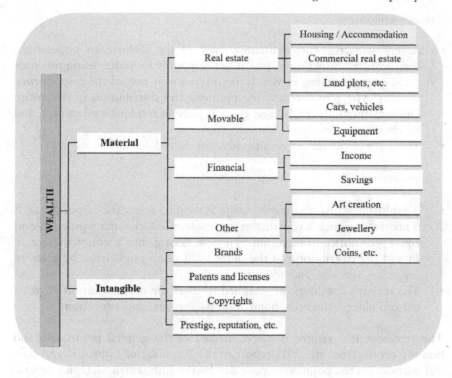

Figure 2.2 A classification of wealth.

Source: Compiled by the author.

method of the Lithuanian population is conducted at the respondent's home, using pre-prepared questionnaires. In this way, the aim is to identify Lithuania's economic inequality (income, consumption, and material living conditions) and its impact on the country's socio-economic progress: economic development and the quality of life of the population. The obtained results reflect the opinions of the entire population of Lithuania and the distribution according to age, sex, place of residence, education, and purchasing power. The study population consists of Lithuanian residents aged 18 and older. The sample size is 1,001 permanent residents of Lithuania.

The aim of the study: To investigate the factors determining Lithuanian socio-economic inequality, socio-economic (income, consumption, and material living conditions) inequality of various groups of Lithuanian society and its consequences for the country's economic and social development and quality of life.

Research group: Residents of Lithuania aged 18 and older.
Sample size: at least 1,000 permanent residents of Lithuania.

Research method:

- A representative questionnaire survey of the Lithuanian population, conducted by direct interview at the respondent's home, using pre-prepared questionnaires, in which the interviewers record the respondents' answers. The results reflect the opinions and distribution of the entire Lithuanian population by age, gender, place of residence, education, and purchasing power.
- The error of the survey results must not exceed 3.1 %.

Selection method:

- A multistage, probabilistic selection is used to select the respondents. It is prepared in such a way that every resident of Lithuania would have an equal probability of being interviewed, taking into account the age of 18 and the distribution of the older Lithuanian population by place of residence, age, sex, and education.
- The survey's data must be weighed to meet the distribution of 18 years old and older Lithuanian population by age, sex, and education.

The representative sample is based on data of the general population and housing census from the 2011 report of the Republic of Lithuania and official statistics on the population provided by the Lithuanian Statistics department. Based on this information, the sample proportions are determined in individual municipalities. In further sampling methods, the proportions of large cities, district centers, towns, and rural areas by official population are determined. Determining the proportions of the sample by place of residence ensures that the survey will be conducted in all Lithuanian municipalities and that the distribution of respondents will correspond to the distribution of the country's population according to the criteria of the place of residence.

Used methods. Differentiation can be estimated using the Gini coefficient, which expresses the relationship between the area of a figure bounded by a line (evenly distributed line), a concentration curve (Lorenz curve), and the area of a triangle below the line of uniformity. This coefficient G is generally calculated using the formula:

$$G = 1 - 2\sum_{i=1}^{k} d_{xi}d_{yi}^{K} + \sum_{i=1}^{k} d_{xi}d_{yi},$$

here – d_{xi} – total volume (population) of the total part of the group i;

d_{yi} – part of the group i of the total scope of the attribute (level of income, consumption, assets, and material conditions);

d_{yi}^{K} – the cumulative (accumulated) part of the total volume of the characteristic of the group i; and

k – number of groups.

If the entire study group is divided into ten-level groups and the frequencies are expressed as a percentage, the Gini coefficient is calculated as follows:

$$G = 110 - 0,2 \sum_{i=1}^{k} d_{yi}^{K},$$

here d_{yi}^{K}– the cumulative frequencies in percent.

If the entire study group is divided into five equal parts, the Gini coefficient is calculated as follows:

$$G = 120 - 0,4 \sum_{i=1}^{k} d_{yi}^{K};$$

This coefficient is most commonly used as a measure of inequality in the distribution of traits. The Gini coefficient varies from zero to one.

Coefficients of concentration and differentiation can be calculated to accurately estimate variation in income, consumption, assets, and material conditions. The following formulas are used to calculate them:

$$K_k = \frac{\sum_{i=1}^{n} (f_i - \Phi_i)}{2 \star 100},$$

$$K_d = \frac{\sum_{i=1}^{n} (f_i - \Phi_i)}{\sum_{i=1}^{n} f_i},$$

here f_i– household comparative shares, in percent;
  Φ_i– comparative shares of income, consumption, assets, and material conditions of each group of households, in percent; and
  K_k, K_d – concentration and differentiation factors.

Comparison of edge deciles. The inequality of the income distribution of the population can be assessed both at a certain point in time (e.g., on average in a certain year) and its dynamics can be analyzed over time. Two values (absolute {limit} and relative {decile differentiation coefficient}) are usually calculated to determine the differences between specific deciles. Assessing the average monetary difference between the poorest and richest tenths of the population reveals the extent to which their situation differs (which also leads to differences in consumption patterns), as well as observing whether these differences are increasing (decreasing) through ongoing research. The increase in differentiation is mainly influenced by the growth of income in the upper ninth, tenth (IX, X) deciles. It is also expedient to calculate the differences between adjacent deciles, the relative difference between the first (I) and tenth (X) deciles, to assess whether the so-called social patience threshold is approaching (after which negative processes in the political and social life of the state begin to appear).

Income differences for larger groups can also be calculated. The differences show that the removal of marginal deciles significantly reduces income variation immediately, and further narrowing of the group no longer results in such fundamental changes as the initial one.

When analyzing the income structure, it is worth taking into account the accumulated decile income. For example, when the economic inequality situation moves in a negative direction, the first (I) decile has relatively less and less income, and X has more and more income. Account must also be taken of the fact that the income of the poorest depends most on social benefits, the amount of old-age pensions, and income from work makes up a much smaller proportion – the poorest depend on the redistribution system mostly.

2.4 Model for assessing the impact of excessive inequality on socio-economic progress

2.4.1 *Indicators for measuring economic inequality and socio-economic progress*

When formulating the model for assessing the impact of economic inequality on socio-economic progress, it is taken into account that the concept of socio-economic progress includes three components – economic growth, quality of life, and sustainability. The model is formed in stages: (1) the impact of economic inequality on economic growth is assessed, (2) the impact of economic inequality on quality of life is analyzed, and (3) the impact of economic inequality on sustainability is identified.

Indicators of economic inequality. An analysis of empirical research assessing the impact of economic inequality on socio-economic progress has revealed that researchers tend to study the impact of income inequality on economic growth. Commonly used measures of inequality are Gini coefficient, unequal distribution of income among the richest and poorest members of society by deciles or quintiles, as well as income of the middle group, and poverty level. However, empirical studies find and use less common measures of inequality – inequality in access to education and health services, EHII2008 inequality measurement index, inequality in basic and advanced opportunities used by the UN.

In order to assess the impact of economic inequality on socio-economic progress and to ensure the reliability of the survey results, the empirical study uses five indicators of economic inequality (see Table 2.2):

Table 2.2 Indicators used in empirical research to measure economic inequality

No.	Indicators of economic inequality	Years	Source
1.	Gini coefficient of equalized disposable income (Gini 1)	2010–2019	Eurostat (2020)
2.	Gini coefficient of equalized disposable income before social transfers, when pensions included in social transfers (Gini 2)	2010–2019	Eurostat (2020)
3.	Gini coefficient of equalized disposable income before social transfers, when pensions excluded from social transfers (Gini 3)	2010–2019	Eurostat (2020)
4.	S80/S20 – income quintile ratio (S80/S20)	2010–2019	Eurostat (2020)
5.	X/I deciles' differentiation coefficient (DK arba X/I)	2010–2019	Authors' calculations based on Eurostat (2020)

Source: Compiled by the author.

1 Gini coefficient of equalized disposable income (Gini 1);
2 Gini coefficient of equalized disposable income before social transfers, when pensions included in social transfers (Gini 2);
3 Gini coefficient of equalized disposable income before social transfers, when pensions excluded from social transfers (Gini 3);
4 X/I deciles' differentiation coefficient (DK or X/I); and
5 S80/S20 – income quintile ratio.

Three Gini coefficients were chosen for the specification highlighting the inequality of income distribution:

- *Gini coefficient of equalized disposable income (Gini 1)* shows the inequality of total disposable income. This indicator includes all personal income, including from various sources and from state social support. The indicator highlights the degree of economic inequality following the redistribution of public money and the implementation of social security policies.
- *Gini coefficient of equalized disposable income before social transfers, when pensions included in social transfers (Gini 2)*, shows the real level of economic inequality versus state social assistance for the most vulnerable members of society, including pensioners.
- *Gini coefficient of equalized disposable income before social transfers, when pensions excluded from social transfers (Gini 3)*, highlights the level of economic inequality versus state social assistance to the most vulnerable members of society, but seniors' pensions are included in personal income. The inclusion of pensions in the calculation of economic inequality reduces the level of the latter.

Given that Gini coefficients poorly reveal income differences at the upper and lower edges of the income distribution, the empirical study uses additional indicators of economic inequality, such as the X/I decile differentiation coefficient and the S80/S20 income quintile ratio.

S80/S20 income quintile ratio expressed as the ratio of disposable income between one-fifth (20%) of those with the highest equivalent disposable income (fifth quintile group) and one-fifth of those with the lowest equivalent disposable income (first quintile group). Similar to the S80/S20 income quintile ratio, the X/I decile differentiation coefficient highlights the average difference in monetary units between the tenth (10%) of the poorest and richest populations. As a rule, the increase in differentiation is most often influenced by the growth of income in the upper (IX, X) deciles. It should also be emphasized that the difference between the X and I deciles helps to assess how close the so-called threshold of social patience is approaching, after which negative processes in the political and social life of the state begin to appear.

Indicators of socio-economic progress. Socio-economic progress is a complex concept. As it has been emphasized, socio-economic progress is based on the premise that economic performance indicators do not adequately describe a country's progress. So the concept of socio-economic progress must include both: quality of life and sustainability indicators. Thus, the monograph analyzes socio-economic progress by integrating three areas: economic growth, quality of life, and sustainability.

Economic growth indicators. Economic growth is important for all countries in the world, as it provides opportunities for developed countries to achieve even higher living standards and for developing countries to close the gap with developed countries. Economic growth is influenced by many different factors; their dynamics influence the cyclical fluctuations of the economy.

Until the industrial economy, the most important economic resource was land, in the industrial economy – capital, and in the post-industrial economy knowledge and innovation. When formulating the model for assessing the impact of economic inequality on economic growth in the monograph, the analysis of scientific literature and empirical research performed in the theoretical part is taken into account; therefore, the following economic growth indicators are distinguished and used in the model (see Table 2.3):

1 real GDP per capita (EUR);
2 investment (gross fixed capital formation) (% of GDP);
3 the proportion of population with tertiary education level (%);
4 inflation rate (average annual change) (%);
5 general government expenditures (% of GDP);
6 the level of government debt (% of GDP);
7 research and development expenditure (EUR per capita);

Table 2.3 Empirical research uses economic growth indicators

No.	Economic growth indicators	Years	Source
1.	Real GDP per capita (EUR)	2010–2019	Eurostat (2020)
2.	Investments (gross fixed capital formation) (% of GDP)	2010–2019	Eurostat (2020)
3.	The proportion of population with tertiary education level (%)	2010–2019	Eurostat (2020)
4.	Inflation rate (average annual change) (%)	2010–2019	Eurostat (2020)
5.	General government expenditures (% of GDP)	2010–2019	Eurostat (2020)
6.	The level of government debt (% of GDP)	2010–2019	Eurostat (2020)
7.	Research and development expenditure (EUR per capita)	2010–2019	Eurostat (2020)
8.	Corruption perception index	2010–2019	Transparency International (2020)
9.	The quality of democracy	2010–2019	Eurostat (2020)
10.	Trade volume (imports, exports) (% of GDP)	2010–2019	Eurostat (2020)

Source: Compiled by the author.

8 corruption perception index;
9 the quality of democracy; and
10 trade volume (imports, exports) (% of GDP).

In the initial modeling phase, more potential drivers of economic growth were identified (e.g., R&D expenditure, quality of democracy, etc.), but the final model uses only one of the latter – the Corruption Perceptions Index (see Figure 2.3).

According to Stiglitz, Sen, and Fitoussi (2010) and arguing that economic indicators are insufficient to describe the real state and well-being of a country, indicators of quality of life and sustainability are included and analyzed in the model.

Indicators of life quality. In general, quality of life is defined as a comprehensive approach to all aspects of human life, based on the suitability of the external environment for living and the synthesis of the internal environment, controlled by the individual as a holder of rights and freedom (Starkauskiene, 2011). Despite the fact that the majority of researchers agree on the complexity of the concept of quality of life, there is no universally accepted classification of quality of life factors and no consensus on its determinants. Therefore, given the multifaceted nature of the concept of life quality, it is recognized that one or more factors do not fully reflect the substance of quality of life content, leading to an increasing number of different quality of life factors in the scientific literature.

	GDP per capita	Level of investment	Proportion of persons with tertiary education	Inflation rate	General government expenditure	Government debt	Research expenditure	Corruption Perceptions Index	The quality of democracy	Trade volume
GDP per capita	1,0	0,1	0,6	-0,1	0,2	-0,1	0,8	0,7	0,7	0,4
Level of investment	0,1	1,0	0,1	0,2	-0,2	-0,5	0,2	0,2	0,2	0,1
Proportion of persons with tertiary education	0,6	0,1	1,0	-0,2	-0,0	-0,1	0,5	0,7	0,4	0,2
Inflation rate	0,1	0,2	-0,2	1,0	-0,0	-0,2	-0,0	0,0	0,0	0,0
General government expenditure	0,2	-0,2	-0,0	-0,0	1,0	0,4	0,5	0,3	0,2	-0,4
Government debt	-0,1	-0,5	-0,1	-0,2	0,4	1,0	-0,1	-0,2	-0,2	-0,4
Research expenditure	0,8	0,2	0,5	-0,0	0,5	-0,1	1,0	0,9	0,7	0,0
Corruption Perceptions Index	0,7	0,2	0,7	0,0	0,3	-0,2	0,9	1,0	0,8	0,1
The quality of democracy	0,7	0,2	0,4	0,0	0,2	-0,2	0,7	0,8	1,0	0,4
Trade volume	0,4	0,1	0,2	0,0	-0,4	-0,4	0,0	0,1	0,4	1,0

Figure 2.3 Strength of correlations between economic growth factors.

Source: Compiled by the author.

Often, quality of life is seen as the sum of external and internal environmental factors. The results of the assessment of the external quality of life environment reflect the suitability of the environment for quality living, and the internal quality of life environment shows whether an individual has the opportunity to use a favorable external quality of life environment to meet his needs. Such an approach only once again justifies the synergy between quality of life, economic growth, and sustainability, and their complex impact on the well-being of both the country and the individual.

The external features of the quality of life in the external environment are highlighted by the natural, economic (combining state policies), and social areas. As the areas of economics and sustainability (natural environment) are examined separately in the monograph, the model for assessing the impact of economic inequality on quality of life focuses on the social external environment and its criteria – number of population at risk of poverty or social exclusion, unemployment and emigration, government expenditure order, and security – and their potential impact on socio-economic progress.

The social environment is an important dimension in assessing the quality of life: living and working conditions, access to education and health care, social equality, people, and organizations with which it interacts (family, friends, and belonging to various communities) directly affect a person's quality of life (Starkauskiene, 2011). In this context, social protection plays an equally important role in ensuring the quality of life of a person as much as possible and in reducing the burden when he or she is exposed to the risk of disability, old age, widowhood, and unemployment. For example, poverty can not only lead to the exclusion of people in social life, but also to prevent life satisfaction and quality of life. The main goal of social protection as a measure of state policy is to ensure the quality of life of an individual and to preserve social harmony, focusing on the principles of security and equality. Studies show that social protection to reduce the risks posed by the aforementioned factors is directly proportional to the level of quality of life in the country.

The theory of possibilities of Sen (1993, 1999) is based on distinguishing and substantiating the factors that determine the high internal quality of life of an individual, when the main emphasis in assessing a person's quality of life is human possibilities – what he can do and what he can be. The features of the internal environment are more emphasized by the complex of physical and personal development well-being, material, and social well-being:

- Health status and personal safety are the foundation of physical well-being (Sen, 1999). Global research substantiates the direct link between the health status of the population and quality of life. When analyzing the health factor in the context of quality of life, a distinction should be made between physical and psychological status, as the latter is often associated with the risk of various illnesses, deaths, or even suicides. The monograph evaluates this area in terms of life expectancy and healthy life years.

- The direct dependence of the quality of life on the well-being of personal development should not be raised. Global research shows that people with higher education are more able to adapt to the environment, have a lower risk of unemployment and wider access to their favorite, well-paid jobs, and improve their status in society. Given that the well-being of personal development is a multifaceted concept, it was decided to use the Human Development Index, which comprehensively measures the average life expectancy, literacy, education, and living standards of the population of all countries of the world. This index makes it possible to determine the social conditions for the well-being of life and personal development in the state. The monograph evaluates this area in terms of the average duration of learning.
- At the level of an individual's quality of life, material well-being is defined by income, savings, wealth, and housing and its quality. Income and other material goods are resources that expand human capabilities, allow them to function freely and pursue a high quality of life (Sen, 1999). The monograph measures this area as the median of equivalent net income and as an indicator of housing deprivation.

Thus, the formation of the model for assessing the impact of economic inequality on the quality of life in the monograph is based on the analysis of scientific literature and empirical research; therefore, the following indicators are distinguished and the model evaluates the quality of life (see Table 2.4):

- median net equivalent income (EUR);
- life expectancy (years);
- the number of people at risk of poverty or social exclusion (%);
- a healthy life year (years);

Table 2.4 Indicators of life quality used in empirical researches

No.	Quality of life indicators	Years	Source
1.	Median net equivalent income (EUR)	2010–2019	Eurostat (2020)
2.	Life expectancy (years)	2010–2019	Eurostat (2020)
3.	The number of people at risk of poverty or social exclusion (%)	2010–2019	Eurostat (2020)
4.	A healthy life year (years)	2010–2019	Eurostat (2020)
5.	The average expected years of schooling (years)	2010–2019	Eurostat (2020)
6.	Unemployment rate (% of the total population)	2010–2019	Eurostat (2020)
7.	Emigration rate (units)	2010–2019	Eurostat (2020)
8.	Severe housing deprivation rate (%)	2010–2019	Eurostat (2020)
9.	Government spending on public order and safety (% of GDP)	2010–2019	Eurostat (2020)

Source: Compiled by the author.

- the average expected years of schooling;
- unemployment rate (% of the total population);
- emigration rate (units);
- severe housing deprivation rate (%); and
- government spending on public order and safety (% of GDP).

By stating that overall socio-economic progress and the quality of life of humans and future generations are unimaginable without a balanced quality of the natural environment, sustainability indicators are distinguished in the next stage of modeling the impact of economic inequality on socio-economic progress.

Sustainability indicators. The pursuit of socio-economic progress is inseparable from sustainability, as progress and economic growth must be reconciled with other social and environmental goals that are important to society.

Despite the fact that economic growth is the goal of each country, the negative consequences of economic growth – depletion of natural resources and environmental pollution – are increasingly being addressed. Economic growth is closely linked to the development of production, which requires natural resources in addition to capital and labor. Demand for non-renewable natural resources is increasing, although supply is declining; and improper and large-scale use of renewable resources can also eventually run out. In addition, international research substantiates a strong link between productivity growth and environmental pollution. Increasing human impact on nature, water and air pollution, and deforestation pose a very serious threat to life worldwide.

In developing the model for assessing the impact of economic inequality on sustainability in the monograph, the model uses the following factors that may affect sustainability (see Table 2.5):

- forest area (% of the total area of the country);
- government expenditure on the environmental protection (% of GDP);
- fossil fuel energy consumption (% of total resources consumed in the country);
- level of economic development (GDP per capita in EUR);
- per capita consumption (total costs) (EUR);
- investment (gross fixed capital formation) (% of GDP);
- trade volume (imports, exports) (% of GDP); and
- corruption perception index.

In summary, it can be stated that in formulating the model of assessing the impact of economic inequality on socio-economic progress, the complexity of the concept of socio-economic progress has been taken into account, distinguishing the impact of economic inequality on economic growth, quality of life, and sustainability.

Table 2.5 Sustainability factors used in empirical research

No.	Sustainability factors	Years	Source
1.	Forest area (% of the total area of the country)	2010–2016	The World Bank (2020)
2.	Government expenditure on the environmental protection (% of GDP)	2010–2019	Eurostat (2020)
3.	Fossil fuel energy consumption (% of total resources consumed in the country)	2010–2015	The World Bank (2020)
4.	Level of economic development (GDP per capita in EUR)	2010–2019	Eurostat (2020)
5.	Per capita consumption (total costs) (EUR)	2010–2019	Eurostat (2020)
6.	Investment (gross fixed capital formation) (% of GDP)	2010–2019	Eurostat (2020)
7.	Trade volume (imports, exports) (% of GDP)	2010–2019	Eurostat (2020)
8.	Corruption perception index	2010–2019	Transparency International (2020)

Source: Compiled by the author.

2.4.2 Assumptions of empirical research

In order to assess the impact of economic inequality on socio-economic progress, the following hypotheses have been formulated. Procedures for testing each hypothesis are also provided.

- *The stronger impact of economic inequality on economic growth is felt in the longer term.*

An empirical study uses an indicator of the change in real GDP per capita (EUR) to measure economic growth. Assuming that a stronger impact of economic inequality on economic growth could occur in the longer term, not only a 1-year economic growth rate but also an average of 2-, 3-, 4-, and 5-year economic growth rates were chosen to measure economic growth (a dependent variable).

- *Changes in economic inequality have different effects on socio-economic progress (economic growth, quality of life, and sustainability) in groups of countries with different living standards.*

Assuming that the impact of economic inequality on socio-economic progress (economic growth, quality of life, and sustainability) could vary across groups of countries by a standard of living, the EU-28 is divided into: (1) higher living countries with a GDP per capita above and (2) countries with a lower standard of living with a GDP per capita measured in purchasing power standards (PPS) of less than 24,750 EUR. The size of 24,750 EUR is the median GDP per capita of the total sample of the empirical study according to the PPS indicator, so the EU-28 is divided into two equal groups.

The Chow test is performed on the latter two groups of countries to test the assumptions. In the Chow test, the null hypothesis about the stability of the coefficients of the equation of the constructed model is tested, which allows identifying whether the dependencies in the different observation samples are the same. In this way, a regression model is created for each group of countries (sample) and it is checked whether the two models overlap. The idea of the Chow test is that the model equation is estimated for the first part of the observations n_1. This equation is used to predict the remaining values of n_2 observations. The null hypothesis is then tested to assess whether the mean of the residual errors of the predicted values is zero. The null hypothesis in this test is formulated as follows: whether the model estimates are identical in both the evaluation sample and the control sample (hypothesis (H0) is tested based on the Chow test: the effect of factors does not differ in the two samples). The Chow test is implemented in the following sequence:

- for the observation sample n_1, the model equation is evaluated and the sum of the squares of the residual errors (RSS_1) is calculated. Number of degrees of freedom $df = n_1 - k$;
- for the observation sample n_2, the model equation is estimated and the sum of the squares of the residual errors (RSS_2) is calculated. Number of degrees of freedom $df = n_1 - k$;
- for the sample of observations ($n_1 + n_2$), the model equation is evaluated and the sum of squares of residual errors (RSS) is calculated with the number of degrees of freedom ($n_1 + n_2 - k$);

- the value of the F statistic is calculated $F \dfrac{\dfrac{RSS - RSS_1 - RSS_2}{k+1}}{\dfrac{RSS_1 + RSS_2}{\left(n_1 + n_2 - 2k - 2\right)}}$

- The condition of the null hypothesis is tested $F^{H0} \sim F_{1-\alpha}\,(k + 1;\ n_1 + n_2 - 2k - 2)$.

Normal economic inequality affects socio-economic progress in one direction; and when it reaches a breaking point (when economic inequality becomes excessive), socio-economic progress begins to act in the opposite direction (inverted "U" curve).

The impact assessment of economic inequality on socio-economic progress assumes that economic inequality affects socio-economic progress (economic growth, quality of life,. and sustainability) in one direction and changes direction when it reaches a certain turning point. The breaking point is calculated by the formula:

$$LT = e^{-\beta_1 / (2 \, * \, \beta_2)} \text{ or } LT = exp\{-\beta_1 / (2 \star \beta_2)\},$$

when

β_1 – coefficient estimate for economic inequality and
β_2 – coefficient estimate at the square of economic inequality.

2.4.3 Methods for assessing the impact of excessive inequality on economic growth, quality of life, and sustainability

An empirical study to assess the impact of economic inequality on socio-economic progress uses panel data from the EU-28 covering the period of 2008–2019. The survey is based on data officially published by Eurostat, The World Bank, and Transparency International. Regressive analysis of macroeconomic data is used for empirical study. The assessment of the impact of economic inequality on socio-economic progress is performed with the Excel software package and the Gretl open-source software package for econometric analysis with panel data.

A regression analysis of the panel data for the period 2008–2019 is performed to analyze the impact of economic inequality on socio-economic progress and to test the hypotheses raised. Panel data are chosen for analysis because they allow assessing the state of the analyzed indicators at a certain moment and how the analyzed indicators change over time. In this way, the panel data combines time series and intergroup data – allowing the evaluation of the dimensions of time and place.

Ordinary least squares (OLS) regression analysis was used for panel data analysis. The latter method is one of the most commonly used because it makes it possible to analyze a large sample of independent variables. The estimates obtained by this method are unbiased and have the smallest variances. However, with the least squares model (LSM), errors are possible that should be identified by reliability tests, and the errors found should be corrected by redesigning the model.

Thus, in order to check the validity of the LSM, the empirical study performed:

1 *Wooldridge* test to determine autocorrelation. Based on the Wooldridge test, the hypothesis (H0) is tested: first-order autocorrelation is not ruled out if $p > 0.05$. If the LSM is autocorrelated, then the model is transformed using robust standard error regression.
2 *White* test to determine heteroskedasticity. Hypothesis (H0) is tested according to the White test: the error scattering homoskedastic is not ruled out if $p > 0.05$. If the errors of the LSM are heteroskedastic, then the model is transformed using the heteroskedasticity-corrected method.

3 According to the *Hausman* test, the diagnostics of the panel data are performed and the model is transformed into a fixed-effects model or a random-effects model according to the recommendations. The Hausman test explores the choice between fixed and random effect models. Hypothesis (H0) is tested according to the Hausman test: the estimates of the generalized LSM are matched without p if 0.0>.

The fixed-effects model identifies the effect of factors on the time-dependent variable, assuming that the coefficients of the independent variables are different. The most important advantage of the fixed effect model is the ability to estimate the effects of a variable or group of variables that change over time. Meanwhile, the random effect model is applied when the number of time series is less than the number of observation groups available. In this case, the independent variables do not correlate with each other or with the dependent variable, and their variations are random. The advantage is the possibility to include the values of low-variable or non-variable variables (Baltagi, 2005).

In order to assess whether the appropriate model type has been chosen to assess the impact of economic inequality on socio-economic progress (economic growth, quality of life, and sustainability), R2 (coefficient of determination – the percentage of dependent variable behavior is defined by independent variables) and adjusted R2 there are many variables and a small number of observations. In this way, it is determined whether the variables selected and used in the model, which measure socio-economic progress – economic growth, quality of life, and sustainability – are appropriate.

The significance of the studied variables is determined according to the significance levels: 99.0%, 95.0%, and 90.0%. The highest (99.0%) significance level is indicated by three asterisks; with two stars – 95.0% significance level; one star – 90.0% significance level.

Since the primary indicators analyzed were not stationary and their units of measurement were different, they were all logarithmized (except for the inflation indicator, as it is a dynamic indicator) and differentiated (annual changes of all variables are calculated, therefore the obtained results are interpreted as coefficients of elasticity) (Tamasauskiene et al., 2016).

The following equation is used to develop an empirical model of the impact of economic inequality on economic growth:

$$\Delta\ln\left(BVP_{i,t}\right) = \alpha + \beta_3\ln\left(BVP_{i,t-1}\right) + \beta_4\Delta\ln\left(INV_{i,t}\right) + \beta_5\Delta\ln\left(AI_{i,t}\right)$$
$$+ \beta_6\left(IL_{i,t}\right) + \beta_7\Delta\ln\left(VSI_{i,t}\right) + \beta_8\Delta\ln\left(VS_{i,t}\right) + \beta_9\Delta\ln\left(KRP_{i,t}\right)$$
$$+ \beta_{10}\Delta\ln\left(PRK_{i,t}\right) + \beta_1\left(EKN_{i,t}\right) + \beta_2\left(EKN\text{sq}_{i,t}\right) + T_1 2010$$
$$+ \ldots + T_{10} 2019 + \Delta u_{i,t}$$

and

$$
\begin{aligned}
EA_{i,t} = \alpha &+ \beta_3 \ln\left(BVP_{i,t-1}\right) + \beta_4 \Delta\ln\left(INV_{i,t}\right) + \beta_5 \Delta\ln\left(AI_{i,t}\right) + \beta_6\left(IL_{i,t}\right) \\
&+ \beta_7 \Delta\ln\left(VSI_{i,t}\right) + \beta_8 \Delta\ln\left(VS_{i,t}\right) + \beta_9 \Delta\ln\left(KRP_{i,t}\right) + \beta_{10}\Delta\ln\left(PRK_{i,t}\right) \\
&+ \beta_1\left(EKN_{i,t}\right) + \beta_2\left(EKNsq_{i,t}\right) + T_1 2010 + \ldots + T_{10} 2019 + \Delta u_{i,t}
\end{aligned}
$$

where:

α – constant;

β – coefficients indicating how strongly and in what direction the independent variables affect the dependent variable;

$EA_{i,t}$ – economic growth in the country i during the time t is calculated $\Delta\ln(BVP_{i,t})$;

$BVP_{i,t}$ – real GDP per capita (EUR) in the country i during the time t;

$INV_{i,t}$ – investment level indicator in the country i during the time t;

$AI_{i,t}$ – proportion of population with tertiary education level in the country i during the time t;

$IL_{i,t}$ – inflation rate in the country i during the time t;

$VSI_{i,t}$ – general government expenditures in the country i during the time t;

$VS_{i,t}$ – the level of government debt in the country i during the time t;

$KRP_{i,t}$ – corruption perception in the country i during the time t;

$EKN_{i,t}$ – indicator of economic inequality in the country i during the time t;

$EKNsq_{i,t}$ – square of the economic inequality indicator in the country i during the time t;

T_t – time pseudo-variables that absorb the cyclical fluctuations of the dependent variable that are common to all countries studied together; and

$\Delta u_{i,t}$ – idiosyncratic (time-varying) model error.

An empirical model of the impact of economic inequality on quality of life uses the following equation:

$$
\begin{aligned}
\Delta\ln\left(GK_{i,t}\right) = \alpha &+ \beta_3 \Delta\ln\left(VGT_{i,t}\right) + \beta_4 \Delta\ln\left(SKL_{i,t}\right) + \beta_5 \Delta\ln\left(SGM_{i,t}\right) \\
&+ \beta_6 \Delta\ln\left(VMT_{i,t}\right) + \beta_7 \Delta\ln\left(NDL_{i,t}\right) + \beta_8 \Delta\ln\left(EML_{i,t}\right) \\
&+ \beta_9 \Delta\ln\left(BNR_{i,t}\right) + \beta_{10}\Delta\ln\left(VITS_{i,t}\right) + \beta_1\left(EKN_{i,t}\right) \\
&+ \beta_2\left(EKNsq_{i,t}\right) + T_1 2010 + \ldots + T_{10} 2019 + u_{i,t}
\end{aligned}
$$

where:

α – constant;

β – coefficients indicating how strongly and in what direction the independent variables affect the dependent variable;

$GK_{i,t}$ – quality of life in the country i during the time t;

$VGT_{i,t}$ – life expectancy in the country i during the time t;

$SKL_{i,t}$ – the number of people at risk of poverty or social exclusion in the country i during the time t;

$SGM_{i,t}$ – healthy life year in the country i during the time t;

$VMT_{i,t}$ – the average expected years of schooling in the country i during the time t;

$NDL_{i,t}$ – unemployment rate in the country i during the time t;

$EML_{i,t}$ – emigration rate in the country i during the time t;

$BNR_{i,t}$ – severe housing deprivation rate in the country i during the time t;

$VITS_{i,t}$ – government spending on public order and safety in the country i during the time t;

$EKN_{i,t}$ – indicator of economic inequality in the country i during the time t;

$EKNsq_{i,t}$ – square of the economic inequality indicator in the country i during the time t;

T_t – time pseudo-variables that absorb the cyclical fluctuations of the dependent variable that are common to all countries studied together; and

$\Delta u_{i,t}$ – idiosyncratic (time-varying) model error.

An empirical model for the impact of economic inequality on sustainability uses the following equation:

$$\Delta\ln\left(TV_{i,t}\right) = \alpha + \beta_3\Delta\ln\left(APL_{i,t}\right) + \beta_4\Delta\ln\left(IKK_{i,t}\right) + \beta_5\Delta\ln\left(BVP_{i,t}\right)$$
$$+ \beta_6\Delta\ln\left(VRT_{i,t}\right) + \beta_7\Delta\ln\left(INV_{i,t}\right) + \beta_8\Delta\ln\left(PRK_{i,t}\right)$$
$$+ \beta_9\Delta\ln\left(KRP_{i,t}\right) + \beta_1\left(EKN_{i,t}\right) + \beta_2 s\left(EKNsq_{i,t}\right)$$
$$+ T_1 2010 + \ldots + T_{10} 2019 + u_{i,t}$$

where:

α – constant;

β – coefficients indicating how strongly and in what direction the independent variables affect the dependent variable;

$TV_{i,t}$ – sustainability level in the country i during the time t;

$APL_{i,t}$ – government expenditure on the environmental protection in the country i during the time t;

$IKK_{i,t}$ – fossil fuel energy consumption in the country i during the time t;

$BVP_{i,t}$ – level of economic development in the country i during the time t;

$VRT_{i,t}$ – per capita consumption in the country i during the time t;

$INV_{i,t}$ – investment (gross fixed capital formation) in the country i during the time t;

$PRK_{i,t}$ – trade volume (imports, exports) in the country i during the time t;

$KRP_{i,t}$ – corruption perception index in the country i during the time t;

$EKN_{i,t}$ – indicator of economic inequality in the country i during the time t;

$EKNsq_{i,t}$ – square of the economic inequality indicator in the country i during the time t;

T_t – time pseudo-variables that absorb the cyclical fluctuations of the dependent variable that are common to all countries studied together; and

$\Delta u_{i,t}$ – idiosyncratic (time-varying) model error.

In summary, the concept of economic inequality is multidimensional, characterized by an integrated interaction of factors, a complex typology, and a specific impact on socio-economic progress, and therefore the impact of economic inequality on socio-economic progress is not unambiguously defined, as evidenced by research diversity. In this monograph, special emphasis is placed on a differentiated approach and the distinction between justified (normal) and unjustified (excessive) inequality, the unequal distribution of income among groups in society, and the assessment of socio-economic progress. To achieve the validity of economic inequality and its impact on socio-economic progress: (1) the impact of economic inequality on economic growth is analyzed; (2) the impact of economic inequality on quality of life is assessed; and (3) the impact of economic inequality on sustainability is identified.

3 ASSESSMENT OF THE IMPACT OF EXCESSIVE INEQUALITY ON SOCIO-ECONOMIC PROGRESS

3.1 Income inequality in EU countries

In 2015, 193 countries of the world, including Lithuania, signed for achievement of UN Sustainable Development Goals (SDGs) aimed to eradicate poverty, ensure peace, the social and economic well-being of all the world's people, to protect the planet, and combat climate change by 2030. It should be emphasized that the UN SDGs include the reduction of inequality within and between countries (SDG 10), where specific challenges have been identified: (1) "by 2030 to empower all persons and to promote their social, economic and political inclusion, regardless of their age, sex, disability, race, ethnicity, origin, religion, economic or other status"; (2) "ensuring equal opportunities and reducing income inequality, including the elimination of discriminatory laws, policies and practices, and ensure the promotion of appropriate laws, policies and actions"; etc. (United Nations Department of Economic and Social Affairs, 2020)).

The EU takes seriously contribution to the UN's SDGs. One of the key ideas of the EU is the well-being of members of society, based on respect for fundamental individual rights and the principles of equality. Unfortunately, in the context of inequality, there has been no significant change in EU countries over the last ten years. Economic inequality remains a particularly relevant problem in the EU countries, which is especially evident when inequality in countries is measured before social transfers, i.e., before state social assistance to the most vulnerable members of society.

In Figure 3.1 dynamics of Gini coefficient in the EU-28 countries during 2010–2019 period is given.

In the EU as a whole, inequality, measured by the Gini coefficient of equivalized disposable income, has increased by 0.7% over a ten-year period (from 30.5% in 2010 to 30.7% in 2019); however, in many countries the variability of inequality is much more pronounced. In addition, although the Gini coefficient of equivalized disposable income before social transfers when pensions are excluded from social transfers has fallen by 1.1% in the EU over the last ten years; however, the Gini coefficient of equivalized disposable income before social transfers when pensions included in social transfers has

DOI: 10.4324/b22984-4

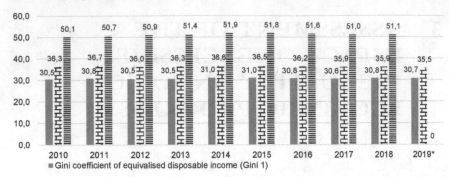

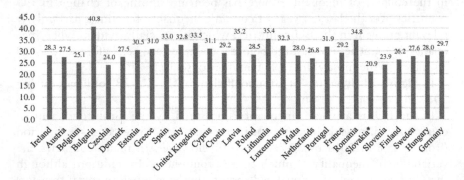

Figure 3.1 Dynamics of Gini coefficient in the EU-28 countries during 2010–2019 period.

* Gini 2 indicator date for 2019 is not available.

Source: Eurostat (2020).

increased by 2%. Thus, although the inclusion of pensions in the calculation of economic inequality reduces the level of the latter, inequality before social transfers remain extremely high in the EU.

According to Eurostat data (2020), in 2019 the fluctuations of the Gini coefficient of equivalized disposable income in the EU countries are shown in Figure 3.2. In 2019, the highest Gini coefficient was recorded in Bulgaria (40.8), Lithuania (35.4), Latvia (35.2), and Romania (34.8). The lowest Gini coefficient was in Finland (26.2), Belgium (25.1), the Czech Republic (24.0), Slovenia (23.9), and Slovakia (20.9).

Figure 3.2 Gini coefficient of equivalized disposable income (2019).

* United Kingdom and Slovakia data in 2018.

Source: Eurostat (2020).

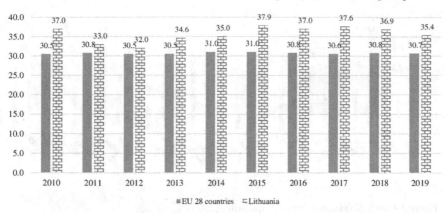

Figure 3.3 Dynamics of Gini coefficient of equivalized disposable income in Lithuania and EU-28 countries in 2010–2019.

Source: Eurostat (2020).

Analysis of dynamics economic inequality in Lithuania and the EU-28, measured by the Gini coefficient of equivalized disposable income showed that over the last ten years the difference between Lithuania and EU-28 average have narrowed (Figure 3.3). In 2010, economic inequality in Lithuania was higher than the average economic inequality of the EU countries by 21%, in 2015 – by 22%; in 2017 – by 23%, while in 2019 – by 15%.

Given the fact that Gini coefficients do little to reveal income differences in the upper and lower edges of income distribution, it is appropriate to use additional indicators of economic inequality, such as the S80/S20 income quintile ratio and the X/I decile differentiation coefficient, when analyzing economic inequality.

According to Eurostat (2020), in the EU-28 member states the largest disparities between the fifth quintile (group with the highest disposable income) and the first quintile (group with the lowest equivalent disposable income) were Bulgaria (8.1), Romania (7.1), Latvia (6.8), Lithuania (6.4), and Italy (6.0). The smallest differences between the fifth and first quintiles were in Finland (3.7), Belgium (3.6), Slovenia (3.4), the Czech Republic (3.3), and Slovakia (3.0) (Figure 3.4). In 2019, the EU-28 average S80/S20 ratio of income in quintiles reached the level of 5.09.

According to Eurostat (2020), in the EU-28 in 2019 the biggest differences in incomes between the tenth richest (X decile) and the tenth poorest (I decile) population groups were in Bulgaria (15.2), Romania (13.6), Italy (12.3), Lithuania (11.5), and Latvia (11.3). Meanwhile, the smallest differences in income between these deciles were recorded in Finland (5.4), Belgium (5.3), Slovenia (4.9), the Czech Republic (4.8), and Slovakia (4.5) (Figure 3.5). In 2019, for EU-28 average, the deciles' differentiation coefficient (X/I) was 8.57.

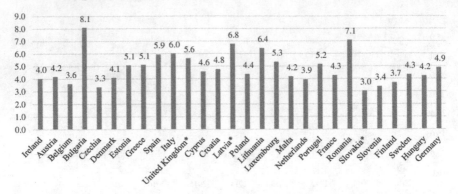

Figure 3.4 S80/S20 income ratio of quintiles (2019).

* United Kingdom, Latvia, and Slovakia data in 2018.

Source: Eurostat (2020).

Attention should be paid to the concentration of the richest decile income, i.e., to X decile income. According to Eurostat (2020), in 2019 in Bulgaria, the richest part of the population (in the 10th decile) accounted for 32% of total income of the country. In other countries, the richest part of population disposed 26.5% of total income in Lithuania; 26.1% – in Cyprus, 26% – in Latvia, and 25.9% – in the United Kingdom (Figure 3.6).

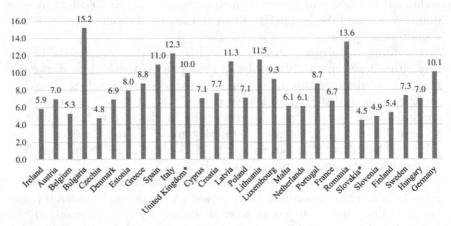

Figure 3.5 X/I decile differentiation coefficient (2019).

* United Kingdom, Latvia, and Slovakia data in 2018.

Source: Eurostat (2020).

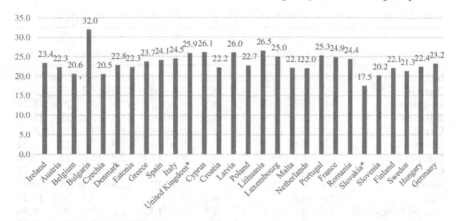

Figure 3.6 Income concentration of the tenth (X) decile in the EU-28 Member States in 2019, % of total income.

* Date for the United Kingdom and Slovakia in 2018

Source: Authors' calculations based on Eurostat (2020).

In Table 3.1, the income distribution by decile in the EU-28 Member States in 2019 is provided.

According to the data provided in Table 3.1, it is clear that the unequal distribution of income among members of society is prevailing in most EU member states. There is a huge income gap between the richest and the poorest; therefore, the concentration of income among the richest people is a serious problem for the whole EU. However, the role of Lithuania as the "anti-leader" is of a great greater concern, as all indicators of inequality are far away from the best performing EU member states.

3.2 Excessive inequality by EU clusters

Assuming that there is an excessive inequality that adversely affects social economic progress, it is particularly important to understand and distinguish between normal (justified, justifiable) inequality and over-inequality or excessive (unjustified, unjustifiable). Normal inequality can have a positive impact on national economic growth, in particular through the promotion of competition and investments, while at the same time creating wide opportunities for the realization of the potential of members of society and the rise of their quality of life. However, when normal inequality is changing into excessive inequality, it can have an opposite or negative impact on the socio-economic progress of countries and can create significant tensions in society, dissatisfaction with quality of life of population, and enhancing vulnerability of people.

Table 3.1 Income distribution by decile in the EU-28 Member States in 2019, %

Countries	I decile	II decile	III decile	IV decile	V decile	VI decile	VII decile	VIII decile	IX decile	X decile	X/I
Ireland	4.0	5.4	6.3	7.2	8.1	9.2	10.4	11.9	14.2	23.4	**5.85**
Austria	3.2	5.5	6.7	7.5	8.5	9.5	10.6	12.0	14.2	22.3	**6.97**
Belgium	3.9	5.6	6.7	7.8	8.8	9.8	10.8	12.1	13.9	20.6	**5.28**
Bulgaria	2.1	3.7	4.8	5.8	6.9	8.3	9.7	11.6	15.1	32.0	**15.24**
Czechia	4.3	6.0	6.9	7.8	8.6	9.5	10.6	11.9	14.0	20.5	**4.77**
Denmark	3.3	5.6	6.7	7.5	8.5	9.4	10.5	11.9	13.8	22.8	**6.91**
Estonia	2.8	4.6	5.7	7.0	8.3	9.6	11.1	12.9	15.6	22.3	**7.96**
Greece	2.7	4.8	6.1	7.1	8.2	9.4	10.8	12.4	14.8	23.7	**8.78**
Spain	2.2	4.5	5.7	6.9	8.1	9.4	10.9	12.8	15.5	24.1	**10.95**
Italy	2.0	4.6	5.9	7.1	8.2	9.4	10.8	12.5	15.1	24.5	**12.25**
United Kingdom*	2.6	4.6	5.7	6.7	7.8	9.0	10.4	12.2	15.1	25.9	**9.96**
Cyprus	3.7	5.0	5.9	6.8	8.0	8.9	10.0	11.5	13.9	26.1	**7.05**
Croatia	2.9	4.9	6.2	7.3	8.7	9.5	10.9	12.6	14.8	22.2	**7.66**
Latvia	2.3	4.1	5.3	6.6	7.8	9.1	10.6	12.7	15.5	26.0	**11.30**
Poland	3.2	5.3	6.4	7.4	8.4	9.4	10.6	12.1	14.5	22.7	**7.09**
Lithuania	2.3	4.2	5.4	6.4	7.6	8.9	10.5	12.6	15.6	26.5	**11.52**
Luxembourg	2.7	4.8	5.9	6.9	7.9	9.1	10.5	12.2	15.0	25.0	**9.26**
Malta	3.6	5.1	6.2	7.3	8.4	9.5	10.7	12.4	14.5	22.1	**6.14**
Netherlands	3.6	5.6	6.7	7.6	8.5	9.5	10.6	12.0	14.1	22.0	**6.11**
Portugal	2.9	4.8	5.9	7.0	8.0	9.0	10.3	11.9	14.8	25.3	**8.72**
France	3.7	5.3	6.4	7.3	8.2	9.0	10.0	11.4	13.7	24.9	**6.73**
Romania	1.8	3.9	5.4	6.7	8.1	9.4	10.9	13.2	16.2	24.4	**13.56**
Slovakia*	3.9	6.3	7.5	8.4	9.2	10.1	11.1	12.3	13.6	17.5	**4.49**
Slovenia	4.1	5.9	7.0	7.9	8.8	9.7	10.7	11.9	13.8	20.2	**4.93**
Finland	4.1	5.6	6.6	7.5	8.4	9.4	10.6	11.9	13.8	22.1	**5.39**
Sweden	2.9	5.3	6.5	7.7	8.8	9.8	11.0	12.4	14.3	21.3	**7.34**
Hungary	3.2	5.5	6.5	7.4	8.4	9.4	10.6	12.0	14.4	22.4	**7.00**
Germany	2.3	5.4	6,5	7.5	8.5	9.5	10.7	12.1	14.4	23.2	**10.09**

* United Kingdom and Slovakia data in 2018.

Source: Authors' calculations based on Eurostat, 2020

Research that explores a differentiated approach to economic inequality through the prism of normal and excessive inequality is rare. Therefore, this monograph aims to distinguish the normal and excessive inequality. Bellow, the main indicators are discussed and empirically tested by applying Eurostat data for EU Member States.

Assessing excessive inequality: Method I. Excessive economic inequality is one of the most sensitive themes in modern society, but the concepts of normal and excessive economic inequality are still emerging. Assuming that *inequality that is considered to be normal in European countries, with coefficient of differentiation between deciles X/I (K_d) lower than 10. It means that* 10% of the richest income exceeds 10% of the poorest income, *is no more than 10 times.*

In Figure 3.7, the differentiation ratio of deciles X/I in the EU-28 in 2019 is given.

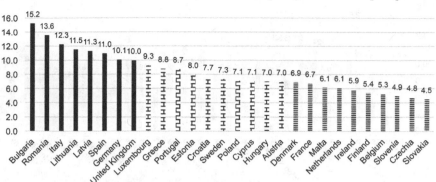

Figure 3.7 X/I deciles' differentiation ratio (2019).

Source: Authors' calculations based on Eurostat (2020).

The X/I deciles' differentiation ratio of the EU-28 countries allows to divide countries into three groups:

1 *Countries with high excessive inequality* ($K_d \geq 10$). Inequality is excessive when the coefficient of differentiation of deciles is more than 10 (or equal 10). The excessive inequality is recorded in Bulgaria ($K_d = 15.2$), Romania ($K_d = 13.6$), Italy ($K_d = 12.3$), Lithuania ($K_d = 11.1$), Latvia ($K_d = 11.3$), Spain ($K_d = 11.0$), Germany ($K_d = 10.1$), and the United Kingdom ($K_d = 10.0$).

2 *Countries with medium and low inequality* ($10 > K_d \leq 7$). The countries with medium or low inequality have the coefficient of differentiation of deciles in the range between 7 and 10. Therefore, inequality can be considered tolerable according to X/I ratio in Luxembourg ($K_d = 9.3$), Greece ($K_d = 8.8$), Portugal ($K_d = 8.7$), Estonia ($K_d = 8.0$), Croatia ($K_d = 7.7$), Sweden ($K_d = 7.3$), Poland ($K_d = 7.1$), Cyprus ($K_d = 7.1$), Hungary ($K_d = 7.0$), and Austria ($K_d = 7.0$).

3 *Countries with normal inequality* ($K_d < 7$). Inequality is normal when the coefficient of differentiation of deciles is less than 7. Countries of normal inequality are: Denmark ($K_d = 6.9$), France ($K_d = 6,7$), Malta ($K_d = 6.1$), The Netherlands ($K_d = 6.1$), Ireland ($K_d = 5.9$), Finland ($K_d = 5.4$), Belgium ($K_d = 5.3$), Slovenia ($K_d = 4.9$), Czech Republic ($K_d = 4.8$), and Slovakia ($K_d = 4.5$).

According to the methodology set out in this monograph, *excessive inequality is one that occurs in the context of an uneven distribution of income among the population, where 10% of the richest population in the country has 10 times higher income then the 10% of the population having lowest earnings in the country. In this case, normal inequality can be achieved if the income of all the poor were increased in such a way that would lead to a more even distribution of income in society, which would affect the decrease in the I/X coefficient to 7.*

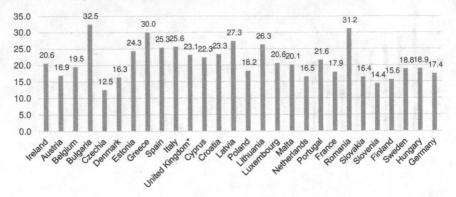

Figure 3.8 At-risk-of-poverty rate in EU-28 in 2019, %.

* United Kingdom data for 2018.

Source: Eurostat (2020).

Assessing excessive inequality: Method II. The concept of economic inequality is closely linked to the concept of poverty. Inequality highlights income distribution, while relative poverty reflects the number of people with low incomes compared to middle-income people. Poverty is linked to economic conditions in which a person cannot afford to participate fully in society's life. The poverty in the EU is sensitive problem and can highlighted by the several facts (Eurostat, 2020). Firstly, in 2019, there were 107.5 million people living in households at risk of poverty or social exclusion in the EU-28. Secondly, in 2019, in more than half of the EU-28 countries at least one fifth of the population was in the group of poverty risk.

In Figure 3.8, at-risk-of-poverty rate in EU-28 in 2019 is provided. According to Eurostat, (2020), at-risk-of-poverty rate (Sk_{real}) in the EU-28 in 2019 was 21.4%. The highest at-risk-of-poverty rate in 2019 was recorded in Bulgaria (32.5%), Romania (31.2%), Greece (30.0%), Latvia (27.3%), and Lithuania (26.3%). The lowest at-risk-of-poverty rates were in Slovakia (16.4%), Denmark (16.3%), Finland (15.5%), Slovenia (14.4%), and the Czech Republic (12.5%).

Assuming that the poverty indicators of Scandinavian countries (Denmark, Finland, etc.) are optimal ($Sk_{optimal}$), it is possible to calculate excessive inequality (PN_{Sk}) on the basis of the formula below, assuming that unjustifiable (excessive) inequality is the difference between the existing (real) poverty level (at-risk-of-poverty rate) and the tolerable (optimal) level of poverty (at-risk-of-poverty rate):

$$PN_{Sk} = Sk_{real} - Sk_{optimal}$$

In Figure 3.9, the excessive inequality in EU countries based on poverty indicators in 2019 is provided.

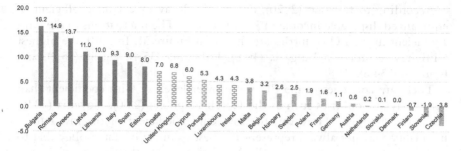

Figure 3.9 Excessive inequality in EU countries based on poverty indicators, 2019.

⋆ United Kingdom data for 2018.

Source: Authors' calculations based on Eurostat (2020).

Three groups of countries emerge from the calculation of the excessive inequality in accordance with the above formula (see Figure 3.9 for data):

1   *Countries with high excessive inequality ($PN_{Sk} > 7$).* Countries with high excessive inequality are: Latvia ($PN_{Sk} = 11.0$), Lithuania ($PN_{Sk} = 10$), Italy ($PN_{Sk} = 9.3$), Spain ($PN_{Sk} = 9.0$), and Estonia ($PN_{Sk} = 8.0$).

2   *Countries of medium and low inequality ($7 \geq K_d \geq 4$).* The group of countries with minor excessive inequality includes Croatia ($PN_{Sk} = 7.0$), the United Kingdom ($PN_{Sk} = 6.8$), Cyprus ($PN_{Sk} = 6.0$), Portugal ($PN_{Sk} = 5.3$), Luxembourg ($PN_{Sk} = 4.3$), and Ireland ($PN_{Sk} = 4.3$).

3   *Countries of normal inequality ($PN_{Sk} < 4$).* According to the latter methodology, inequality is normal in Malta ($PN_{Sk} = 3.8$), Belgium ($PN_{Sk} = 3.2$), Hungary ($PN_{Sk} = 2.6$), Sweden ($PN_{Sk} = 2.5$), Poland ($PN_{Sk} = 1.9$), France ($PN_{Sk} = 1.6$), Germany ($PN_{Sk} = 1.1$), Austria ($PN_{Sk} = 0.6$), Netherlands ($PN_{Sk} = 0.2$), Slovakia ($PN_{Sk} = 0.1$), Denmark ($PN_{Sk} = 0$), Finland ($PN_{Sk} = 0.7$), Slovenia ($PN_{Sk} = -1.9$), and the Czech Republic ($PN_{Sk} = 3.8$).

According to the methodology set out in this monograph, *normal and excessive inequality are determined using the optimal poverty threshold.* In this case, *normal inequality is part of the overall inequality corresponding to poverty-free inequality, i.e., when the country's at-risk-of-poverty rate does not exceed the range from 15.6% to 16.5%.*

Assessing excessive inequality: Method III. The Gini coefficient is one of the most commonly used statistical indicators of the distribution of income inequality; so, the latter could also be adapted to the isolation of excessive inequality. In this way, *excessive inequality is interpreted as overcoming normal or, otherwise, optimal inequality.* In this context, it is assumed that normal/ optimal inequality is best reflected by the Scandinavian countries' Gini coefficient in terms of gross disposable income, which ranged from 26.2 to 27 in 2019.

According to Eurostat (2020), the in EU-28 average Gini coefficient of equivalized disposable income in 2019 was 30.7. The fluctuations in the Gini coefficient in the EU countries are shown in Figure 51. In 2019, the highest Gini was recorded in Bulgaria (40.8), Lithuania (35.4), Latvia (35.2), and Romania (34.8).

There are countries in the EU with lower levels of Gini coefficient than Scandinavian countries such as the Czech Republic (24.0), Slovenia (23.9), Slovakia (20.9), etc. However, the low Gini coefficient or the low rate of inequality may not always represent the best example of state policy that to be followed (e.g. socialist state policy). For this reason, the Gini coefficient of the Scandinavian countries is used as a starting point for assessment of normal inequality in this monograph. In this way, excessive inequality (PN_{Gini}) is calculated according to the formula below, assuming that unjustifiable or excessive inequality is the difference between the existing (real) Gini coefficient and the tolerable (optimal) Gini coefficient level:

$$PN_{Gini} = Gini_{real} - Gini_{optimal}$$

In Figure 3.10, the excessive inequality in EU countries by Gini coefficient in 2019 is provided.

According to the calculations carried out, the countries could be divided into three clusters (see Figure 3.10).

1   *Countries with high excessive inequality ($PN_{Gini} > 4$).* Countries of high excessive inequality include Bulgaria ($PN_{Gini} = 13.3$), Lithuania ($PN_{Gini} = 7.9$), Latvia ($PN_{Gini} = 7.7$), Romania ($PN_{Gini} = 7.3$), the United Kingdom ($PN_{Gini} = 6.0$), Spain ($PN_{Gini} = 5.5$), Italy ($PN_{Gini} = 5.3$), Luxembourg ($PN_{Gini} = 4.8$), and Portugal ($PN_{Gini} = 4.4$).

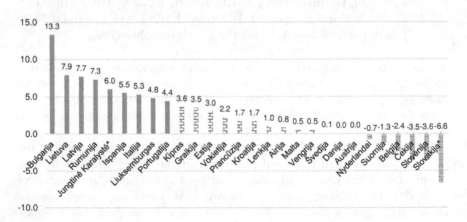

Figure 3.10 Excessive inequality in EU countries by Gini coefficient, 2019.

* United Kingdom and Slovakia data for 2018.

Source: Authors' calculations based on Eurostat (2020).

2 *Medium- and low-inequality countries (4 ≤ PN$_{Gini}$ ≥ 0.1).* The tolerable group of countries according to excessive equality includes Cyprus (PN_{Gini} = 3.6), Greece (PN_{Gini} = 3.5), Estonian (PN_{Gini} = 3.0), Germany (PN_{Gini} = 2.2), France (PN_{Gini} = 1.7), Croatia (PN_{Gini} = 1.7), Poland (PN_{Gini} = 1.0), Ireland (PN_{Gini} = 0.8), Malta (PN_{Gini} = 0.5), Hungary (PN_{Gini} = 0.5), and Sweden (PN_{Gini} = 0.1).

3 *Countries of normal inequality (PN$_{Gini}$ ≤ 0).* According to the latter methodology, there is no excessive inequality in Denmark (PN_{Gini} = 0), Austria (PN_{Gini} = 0), the Netherlands (PN_{Gini} = −0.7), Finland (PN_{Gini} = −1.3), Belgium (PN_{Gini} = −2.4), the Czech Republic (PN_{Gini} = −3.5), Slovenia (PN_{Gini} = −3.6), and Slovakia (PN_{Gini} = −6.6).

A summary of all three methodologies for calculating excessive inequality and the highlighted clusters of three countries is presented in Table 3.2.

Attention should be paid to the countries of the first cluster, i.e., countries of high excessive inequality (Figure 3.11). *In all three cases of methods for calculating excessive inequality, the same states "anti-leaders" are revealed: Bulgaria, Romania, Lithuania, Latvia, Spain, and Italy.*

The results of the research performed allow developing the following conclusions:

• Excessive inequality is one that occurs in the context of an uneven distribution of income among the population, where 10% of the country's richest population has 10 times higher income than the 10% of the country's lowest-earning population. In this case, normal inequality would be the inequality that would arise if the incomes of all the poor were increased in such a way that would lead to a more even distribution of income in the lower deciles and which would affect the decrease of the I/X coefficient of differentiation of deciles income to 7.

• Normal and excessive inequality are identified by using the optimal at-poverty-risk rate. In this case, normal inequality is part of the overall inequality corresponding to poverty-free inequality, i.e., when the country's at-poverty-risk rate does not exceed the interval from 15.6 to 16.5%. Excessive inequality is part of the overall inequality caused by poverty, which exceeds the mentioned optimal poverty threshold.

• Excessive inequality is interpreted as overreaching of general inequality as normal or, otherwise, optimal, i.e., when the Gini coefficient of equivalized disposable income overcomes the range of 26.2 to 27.6.

Economic inequality separated into excessive and normal allows to divide countries into three clusters: (1) countries of high excessive inequality, (2) countries of medium and small inequality, and (3) countries of normal inequality. Based on all three methods of calculation in the first cluster of countries – countries of high excessive inequality there are the same "anti-leaders": Bulgaria, Romania, Lithuania, Latvia, Spain, and Italy.

Table 3.2 Summary of methodologies for calculating excessive inequality

Methodology for calculating excessive inequality (PN)	Group of countries I	Group of countries II	Group of countries III
By coefficient of differentiation of deciles (PNK_d)	($K_d \geq 10$)	($10 > K_d \geq 7$)	($K_d < 7$)
	Bulgaria ($K_d = 15.2$), Romania ($K_d = 13.6$), Italy ($K_d = 12.3$), Lithuania ($K_d = 11.5$), Latvia ($K_d = 11.3$), Spain ($K_d = 11.0$), Germany ($K_d = 10.1$), United Kingdom ($K_d = 10.0$)	Luxembourg ($K_d = 9.3$), Greece ($K_d = 8.8$), Portugal ($K_d = 8.7$), Estonia ($K_d = 8.0$), Croatia ($K_d = 7.7$), Sweden ($K_d = 7.3$), Poland ($K_d = 7.1$), Cyprus ($K_d = 7.1$), Hungary ($K_d = 7.0$), Austria ($K_d = 7.0$)	Denmark ($K_d = 6.9$), France ($K_d = 6.7$), Malta ($K_d = 6.1$), Netherlands ($K_d = 6.1$), Ireland ($K_d = 5.9$), Finland ($K_d = 5.4$), Belgium ($K_d = 5.3$), Slovenia ($K_d = 4.9$), Czech Republic ($K_d = 4.8$), Slovakia ($K_d = 4.5$)
By poverty indicator (PN_{Sk})	($PN_{Sk} > 7$)	($7 \geq PN_{Sk} \geq 4$)	($PN_{Sk} < 4$)
	Bulgaria ($PN_{Sk} = 16.2$), Romania ($PN_{Sk} = 14.9$), Greece ($PN_{Sk} = 13.7$), Latvia ($PN_{Sk} = 11.0$), Lithuania ($PN_{Sk} = 10$), Italy ($PN_{Sk} = 9.3$), Spain ($PN_{Sk} = 9.0$), Estonia ($PN_{Sk} = 8.0$)	Croatia ($PN_{Sk} = 7.0$), United Kingdom ($PN_{Sk} = 6.8$), Cyprus ($PN_{Sk} = 6.0$), Portugal ($PN_{Sk} = 5.3$), Luxembourg ($PN_{Sk} = 4.3$), Ireland ($PN_{Sk} = 4.3$)	Malta ($PN_{Sk} = 3.8$), Belgium ($PN_{Sk} = 3.2$), Hungary ($PN_{Sk} = 2.6$), Sweden ($PN_{Sk} = 2.5$), Poland ($PN_{Sk} = 1.9$), France ($PN_{Sk} = 1.6$), Germany ($PN_{Sk} = 1.1$), Austria ($PN_{Sk} = 0.6$), Netherlands ($PN_{Sk} = 0.2$), Slovakia ($PN_{Sk} = 0.1$), Denmark ($PN_{Sk} = 0$), Finland ($PN_{Sk} = -0.7$), Slovenia ($PN_{Sk} = -1.9$), Czech Republic ($PN_{Sk} = -3.8$)
By Gini coefficient (PN_{Gini})	($PN_{Gini} > 4$)	($PN_{Gini}\ 4 \leq \geq 0.1$)	($PN_{Gini} < 0$)
	Bulgaria ($PN_{Gini} = 13.3$) Lithuania ($PN_{Gini} = 7.9$), Latvia ($PN_{Gini} = 7.7$), Romania ($PN_{Gini} = 7.3$), United Kingdom ($PN_{Gini} = 6.0$), Spain ($PN_{Gini} = 5.5$), Italy ($PN_{Gini} = 5.3$), Luxembourg ($PN_{Gini} = 4.8$), Portugal ($PN_{Gini} = 4.4$)	Cyprus ($PN_{Gini} = 3.6$), Greece ($PN_{Gini} = 3.5$), Estonia ($PN_{Gini} = 3.0$), Germany ($PN_{Gini} = 2.2$), France ($PN_{Gini} = 1.7$), Croatia ($PN_{Gini} = 1.7$), Poland ($PN_{Gini} = 1.0$), Ireland ($PN_{Gini} = 0.8$), Malta ($PN_{Gini} = 0.5$), Hungary ($PN_{Gini} = 0.5$), Sweden ($PN_{Gini} = 0.1$)	Denmark ($PN_{Gini} = 0$), Austria ($PN_{Gini} = 0$), Netherlands ($PN_{Gini} = -0.7$), Finland ($PN_{Gini} = -1.3$), Belgium ($PN_{Gini} = -2.4$), Czech Republic ($PN_{Gini} = -3.5$), Slovenia ($PN_{Gini} = -3.6$), Slovakia ($PN_{Gini} = -6.6$)

Source: Authors' calculations.

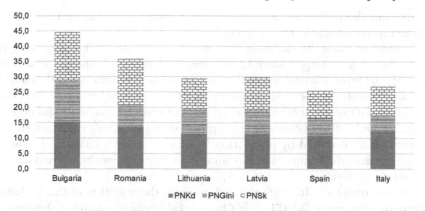

Figure 3.11 Countries with high inequality (according to the methods presented).

Note: PNK$_d$ – Excessive inequality by coefficient of differentiation of deciles PN$_{Sk}$ – excessive inequality by poverty indicators, PN$_{Gini}$ – excessive inequality by Gini coefficient.

Source: Authors' calculations.

Group of countries of normal inequality includes Denmark, Finland, Sweden, the Netherlands, Belgium, Austria, the Czech Republic, Slovenia, and Slovakia.

3.3 Wealth inequality: The case of Lithuania

Possession and management of property or wealth affects the person not only material, but also moral security, self-confidence, self-worth, and the corresponding quality of life. Historically, it the real estate was treated as the most valuable property. House is a particularly important part of human well-being and, consequently, a guarantee of societal stability. However, in the scientific literature, this phenomenon is still under-focused due to lack of data, taking into account the complexity of assessing wealth inequality and the impact of available wealth on the quality of human life.

In 2010, EU countries launched the Europe 2020 strategy for growth and jobs, which aimed to increase the EU's competitiveness and significantly increase resource efficiency. The strategy's main idea is the partnership between EU Member States with priority-oriented objectives and a specific implementation framework. It sets out five interrelated headline targets to be achieved by EU Member States until 2020 in the areas of employment, research and development, climate change mitigation and energy, education and the fight against poverty, and social exclusion (European Commission, 2015). It is argued that one of the main challenges of the Europe 2020 strategy is precisely the reduction of poverty and social exclusion with the objective of providing every EU citizen with decent housing in terms of both price and quality. According to Eurostat, quality of housing is the key to quality of life, but the lack of adequate housing is a longstanding problem in many European countries (Eurostat, 2014).

In order to assess the success in implementing Europe 2020 strategy targets and achievements in providing EU citizens with quality housing and also to identify the level of wealth inequality in Lithuania, the authors jointly carried out a representative sociological survey of Lithuanian residents, to collect the necessary data on the wealth disposal of the population (real estate and other), together with criteria characterizing physical and qualitative characteristics of the wealth disposed by population in Lithuania. This chapter of mono-graph presents the results of the sociological survey on wealth carried out in Lithuania in 2016.

Wealth distribution. In 2016, the authors with the scientists of the Mykolas Romeris University (MRU) Life Quality Laboratory organized the repre-sentative survey of Lithuanian residents (1000 respondents) and used a certain survey in order to collect information about the dwelling-places or other type of property owned by the respondents. The survey was organized on the basis of such wealth-defining parameters as the type of a dwelling-place, its size, construction year, location, etc. Also, questions were asked about the disposal of other assets and its functional type (homestead, garden house, agricultural plot, forest, securities, artwork, jewelry, coins, postage stamps, and other valuable collectibles). These attributes may determine the value of housing or other assets, taking into account market prices at the time of the survey. Respondents were also asked about the property rights and the way in which the property was acquired, i.e., with the loan, without loan, donated, inherited, etc.

According to survey results, 85.2% of respondents live in their own houses (853 respondents), 11.4% (114) of population rent they houses. The 8.7% of population of Lithuania has a second dwelling, 34.9% – own plot of land (more than 5 acres).

According to income, in their own housing live 96.6% retired people; 86.4% – of people having income form from own business or farming and 81.1% of the employed population. A relatively high proportion – 82.3% of those receiving unemployment benefit, disability pension, or other social benefits also have their own housing. Only more than half (60%) of peo-ple living on child benefits do not own housing. More than one dwelling is owned by self-employed or farmers (15.9%) and also employed persons (10.3%), and 7.8% of those living form pensions, as well as 2.5% of unem-ployed persons, receiving disability pensions or other social benefits. The 56.8% of business owners or farmers are land (more than 5 acres) owners. Also, 40.5% of pensioners and 33.5% of employed people and more than a fifth, i.e., 21.5% of the population whose main source of income is unem-ployment benefit, disability pension, or other social benefits are land (more than 5 acres) owners.

According to the survey data, from 0.8% to 26.5% of the respondents (Table 3.3) have disposable property, which, if necessary, could be sold,

Table 3.3 Households with other property, which may be sold, pledged, and leased, if necessary

Type of asset	N	Percent
Homestead	119	11.9
Garden cottage	140	14.0
Agricultural land	265	26.5
Forest	106	10.6
Shares, bonds	41	4.1
Securities	22	2.2
Valuable artwork	18	1.8
Expensive jewelry	65	6.5
Other assets (coin, stamp collections, amber, gold, etc.)	77	7.7
Other: house/apartment (0.4%); garage (0.3%); books etc.	8	0.8

Source: MRU Quality of Life Laboratory Research (2016).

pledged, or leased. Most of these people are employed, self-employed, or farmers and also retired. Residents with particularly low-income levels (those receiving unemployment benefits, disability pensions, or other social benefits) have an agricultural parcel (15.2% of such persons), garden cottages (7.6%), and farmstead 6.3%. Households living on child allowances do not have any property with potential for sale, pledge, or lease.

The pessimistic self-assessment of respondent's situation is also demonstrated by self-classification toward different groups of socio-economic situations: some respondents tend to be sufficiently critical of their material situation, although the available property at their disposal, identified during the survey, tends to allow them to meet their necessary needs.

According to the subjective self-assessment of the socio-economic situation, respondents assigning themselves to the poor, who do not have money for dignified life, have the following wealth accumulated: 90.6% have their own housing, 43.5% have a car (2.2% of them have 2 cars), 33% have their own plot of land (of more than 5 acres); and have the property that could be sold, rented, pledged (9.1% have a farmstead, 14.1% − garden cottage, 22.1% − agricultural land plot, 5.1% − forest, 5.8% − other property such as coins, stamp collection, amber, gold, etc.).

Those who attribute themselves to the poorest group, who do not have money for basic everyday needs, have the following property: 82.2% live in their own housing, 11.1% have a car, and 15.6% have their own plot of land (of more than 5 acres) (Figure 3.12).

While the middle group, who have sufficient money to live with dignity, and middle group who sometimes have to limit costs (do not always have money to live with dignity) usually have their own housing (82.4% and 83.5%, respectively). The most of poorest segment people (those who are poor and can't live with dignity and the poorest people) have their own housing as well (90.6% and 82.2%, respectively). Also, 36% and 37.1% of poor

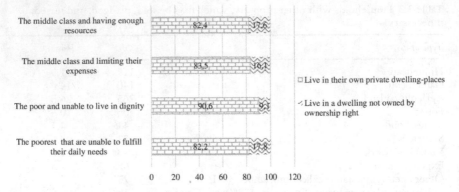

Figure 3.12 Share of population owning housing (%) in different socio-economic situations (by subjective self-stratification).

Source: MRU Quality of Life Laboratory Research (2016).

(those who are poor and can't live with dignity) and the poorest group of people have their own plot of land (home holding) with more than 5 acres.

By residential area, own housing is mainly distributed in towns and cities (91% and 95%, respectively), in villages (84.4%), and metropolitan areas (83.3%). The lowest share of having own housing – 80.3% – is held by the capital's residents (Figure 3.13).

The analysis of the data by age group shows that the acquisition of own housing depends directly on the age of the resident: the older the population, the greater the share of their age group has their own housing (Figure 3.14).

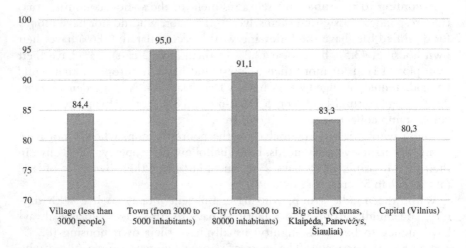

Figure 3.13 Share of residents having own housing (%) by place of residence.

Source: MRU Quality of Life Laboratory Research (2016).

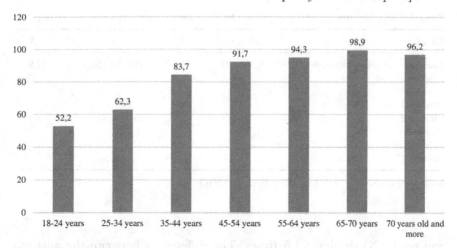

Figure 3.14 Share of population living in own housing (%) by age.

Source: MRU Quality of Life Laboratory Research (2016).

There are no significant differences in the ownership of housing by gender: 85.8% of women surveyed and 84.7% of men live in their own housing.

Assessment of the wealth distribution denominated in property value. For the purpose of determining the wealth distribution measured on the basis of property value, the sample shall include only dwellings acquired without a financial loan and inherited or donated, i.e., the dwellings which are not pledged or otherwise restricted and are wholly owned by the households (598).

In Table 3.5, the distribution of wealth denominated in property value in deciles groups is provided.

According to the data (in Tables 3.4 and Table 3.5) on own housing provided by respondents during the survey, the minimum statistical value per housing, based on property market value, ranges from €9,990 in the first decile (I) to €169,200 in the tenth decile (X); so, the difference is about 16.9 times.

When comparing the average housing prices of the first (I) and tenth (X) deciles, the difference is approximately 7.2 times; the difference between second (II) and ninth (IX) deciles, is 3.5 times; and for adjacent (ninth) IX

Table 3.4 Housing according to the way it is acquired

Of the homeowners/co-owners (N = 723)	N	Percent
Housing purchased without financial loan	472	65.3
Housing purchased with a financial loan	119	16.5
Housing is inherited, donated	126	17.4
Don't know	6	0.8
Total	723	100.0

Source: MRU Quality of Life Laboratory Research (2016).

Table 3.5 Distribution of wealth expressed by the value of property in deciles groups

Deciles	I	II	III	IV	V	VI	VII	VIII	IX	X
Average value of property per decile (€)	13,592	18,468	21,793	26,567	27,260	33,077	38,143	50,026	64,715	97,890
Differences between deciles (times)		1.35	1.18	1.22	1.02	1.21	1.15	1.31	1.29	1.51

Source: MRU Quality of Life Laboratory Research (2016).

and tenth (X) deciles − 1.5 times. The differences between the adjacent deciles are not very big, and they are getting higher then approaching (X) decile (Table 3.6 and Figure 3.15).

Thus, the wealth distribution denominated in the value of housing is sufficiently large, i.e., the differences in the deciles groups between the first (I) and tenth (X) deciles exceeds 7 times. However, it should be noted that this assessment is not sufficiently precise due to the following limitations:

- The wealth was assessed, based on the value of the own housing held by the population, i.e., those without their own housing (representing 14.8% of all respondents) and having purchased housing with a loan (around 12% of respondents) were excluded from the sample of the research. Thus, for the purposes of the analysis and evaluation, only the housing owned by 60% of the respondents has been taken into consideration − no other real estate, agricultural parcels, durable goods, works of art, or other types of assets have been analyzed.
- The value of the main (first) housing was assessed, i.e., the uncollected data did not allow to include in the calculations the value of second dwellings as this survey did not collect detailed data on second dwellings. It only recorded whether the respondent had the second housing. Fifty-two respondents having second houses were recorded. Considering that about half of these dwellings are located in the regions and hypothetically assessing the potential minimum value of second housing of around

Table 3.6 Decile coefficients for assets denominated in housing value

X/I	IX/II	X/V	X/IX
7.2	3.5	3.6	1.5

Source: MRU Quality of Life Laboratory Research (2016).

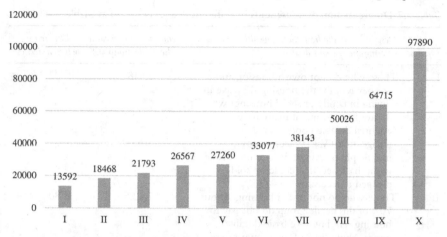

Figure 3.15 Average value of housing in deciles (€).

Source: MRU Quality of Life Laboratory Research (2016).

€20,000, the difference between the first (I) and tenth (X) deciles would increase to 8.7 times.

- The determined housing prices based on survey are only preliminary, to the extent that it allows an objective assessment of the following parameters: the year of construction; the type of housing, the location of housing, etc., because the market prices are not as detailed as the indications set out in the survey (market prices of housing are being provided only for new and old housing).

Thus, the determined deciles' (X/I) differentiation coefficient (7.2 times) only roughly reflects the difference between housing own in the periphery and in metropolitan areas. The adjustment of data and the inclusion of respondents without housing, as well as those at the disposal of second, third, etc., housing and other assets (land, shares, durable goods, etc.) would reveal much more pronounced wealth differences and inequality among the Lithuanian population.

Assessment of the wealth distribution reported by respondents during the survey. The addition of respondents to the research sample who do not own housing, but rent it, have not yet covered the housing loan, or have other assets (e.g., car, other durable goods), as well as second housing, plot of land, and other wealth (agricultural land plots, forests, durable goods, shares, securities, artwork, etc.), provides for more pronounced wealth differences and inequality between the Lithuanian population. According to the survey data, the preliminary assignment of the Lithuanian population to decile groups according to the assets at their disposal is provided in Table 3.7 and Figure 3.16.

Table 3.7 Differentiation of households in different deciles based on available wealth

Deciles	Description of the household's wealth entering the relevant decile	Average value of wealth in the decile group (€)	Difference between deciles
I	Those who do not own a housing and do not pay rent for the housing they live in (living in family or social housing) who do not have a car, but many have essential durable goods such as a refrigerator, TV, gas/electric stove, mobile phone. About 36% of this decile group has assets that could be sold or pledged if necessary.	2,648	
II	Those who do not own a housing. About 70% of this decile group do not pay for housing rent but have basic durable goods. About 90% of them own or are likely to own a car or other property.	7,420	2.80
III	Those who do not own housing but who are able to pay or pay rent for housing. Around 70% of this decile group have a car, about 20% have a plot of land with more than 5 acres or have other real estate.	10,820	1.46
IV	Those having own housing, the value of which, depending on the location (village, city or city), the year of construction, size and type (apartment or individual house, etc.) is about €13–€18,000.	15,180	0.71
V	Those who have acquired their own accommodation with a loan or loan instalments (the proportion of those who didn't pay their loan yet is about 43% of the decile group). About 14% of this decile group has a second housing, and 48% of them – a plot of land greater than 5 acres.	17,683	1.16
VI	Those who have acquired their own accommodation without a loan, the value of which, depending on the location of the dwelling (village, city or city), the year of construction, size and type (apartment or individual house, etc.) is about €18–€25,000.	22,412	1.48
VII	Those who have own dwellings acquired without a loan, the value of which, depending on the location of the dwelling (village, city or city), the year of construction, size and type (apartment or individual house, etc.) is about €25–€29,000.	27,087	1.21

(Continued)

Table 3.7 Differentiation of households in different deciles based on available wealth *(Continued)*

Deciles	Description of the household's wealth entering the relevant decile	Average value of wealth in the decile group (€)	Difference between deciles
VIII	Those who have acquired their own accommodation without a loan, the value of which, depending on the location of the dwelling (village, city or city), the year of construction, size and type (apartment or individual house, etc.) is about €32,000. About 39% of this decile group has a plot of land greater than 5 acres.	38,830	1.43
IX	Those who have acquired their own accommodation without a loan, the value of which, depending on the location of the dwelling (village, city or city), the year of construction, size and type (apartment or individual house, etc.) is about €50,000. This group has also a plot of land greater than 5 acres.	60,261	1.55
X	Those who have acquired their own accommodation without a loan the value of which, depending on the location of the dwelling (village, city, or city), the year of construction, size and type (apartment or individual house, etc.) is about €85,000. This group has also a plot of land greater than 5 acres and 58% of this decile group has a second housing.	108,028	1.79

Source: MRU Quality of Life Laboratory Research (2016).

Based on preliminary estimates, based on the market prices for housing, land, durable goods, and other assets collected in 2016 (at the time of the survey), and the data collected by the survey, it is possible to receive a wealth distribution that significantly outpaces the income and consumption distribution values in Lithuania set and reported by official statistics.

The coefficient of differentiation of deciles (ratio of I and X deciles) is equal to 40.8 (see Table 3.8). It should be noted that according to the survey data, it is not possible to know and objectively estimate the value of the wealth at the disposal of households by indicating their income at intervals by the respondents, and the actual value of the income received by the population at their disposal is not known precisely in this survey.

According to the responses of the richest respondents in the sample, 10% of the richest population in Lithuania are potentially possessing around 34.8% of all Lithuanian wealth and 1% of the richest population – 8.9% of the wealth of Lithuania. Also, when calculating the differences between adjacent decile, it can be sees that the most pronounced distances are

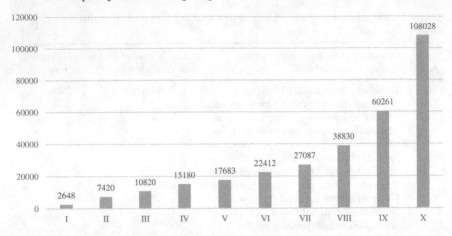

Figure 3.16 Average value of household wealth in deciles (€).

Source: MRU Quality of Life Laboratory Research (2016).

from the tenth (X) and ninth (IX) deciles, and the first (I) and second (II) deciles reaching 2.8 times. This indicates that the value of the wealth of extreme decile groups has a significant impact on wealth inequality rates in Lithuania.

The analysis of own housing acquired with the loan shows the following trends: according to the survey, households living in purchased housing with loans account for 11.9%, of which 72.3% live in flats in apartment buildings. However, the distribution of housing acquired with the loan is slightly different than usual: a significant proportion of housing acquired with a loan are in rural areas and small towns (16% and 22%, respectively) and – 13.2% in cities (from 5,000 to 80,000 inhabitants), 17.7% in big cities such as Kaunas, Klaipėda, Panevėžys, and Šiauliai, and only one fifth of housing acquired with a loan (19.7%) are located in Vilnius. The size of the dwelling is dominated by 60–79 sq. m. (42%) and 50–59 sq. m. (23.5%) of all housing. According to the year of construction, 38.7% is built between 1971 and 1980 and 22.7% between 1981 and 1990. In addition, 21.6% of respondents reported that the repayment of housing loans is a significant burden on the household, and 43.1% households have reported that loan is as a kind of burden for them (Figure 3.17).

Table 3.8 The coefficients of differentiation of deciles for household's assets

X/I	IX/II	X/V	X/IX
40.8	8.1	6.1	1.8

Source: MRU Quality of Life Laboratory Research (2016).

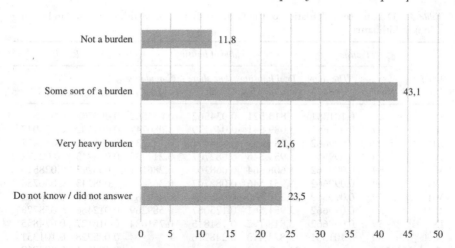

Figure 3.17 Repayment of a housing loan is a burden on this part of households, %.

Source: MRU Quality of Life Laboratory Research (2016).

According to the survey, 11.4% of households rent the dwelling. The distribution of rented housing looks like this: according to the type of rented housing, flats in apartment buildings are dominating (81.6%). More than half of rented housing (54.4%) were built between 1961 and 1980. Two thirds of rented housing (64.1%) are rented by 18–34 years old people. The 66.7% of renters are employed workers whose main source of income is wages. It should be noted that 7.9% of renters are pensioners and 8.8% are households living from unemployment benefit, disability pension, or other social benefits. In assessing the financial capacity of households to rent housing, other costs for maintaining this housing must also be taken into account. The 46.5% of all respondents reported that their spendings on rented housing are between €101 and €200 per month per month. The 34.2% renters reported the spending to €100 and 14% – spending between €201 and €450 per month.

Calculation of the Gini coefficient of the wealth expressed in value of the housing. The Gini coefficient (G) is the ratio between the shape area bounded by the straight line (even distribution of wealth line) and the concentration curve (Lorenz curve) and the triangular area, which is below the line of even distribution of wealth:

$$G = 1 - 2\sum_{i=1}^{k} d_{xi} d_{yi}^{K} \mid \sum_{i=1}^{k} d_{xi} d_{yi}, \tag{1}$$

where
   d_{xi} – the share of total size of the population (in the example, population having 1, 2, rooms apartments) in i-group;

Table 3.9 Data for the calculation of the Gini coefficient of wealth (expressed in housing value) in Lithuania

Deciles No.	Housing		Value of housing			Result	
	Number	The share d_{xi}	Total housing value (€)	The share d_{yi}	Cumulative part d_{yi}^K	$d_{xi}d_{yi}$	$d_{xi}d_{yi}^K$
I	60	0.101351	815,521	0.034562	0.034562	0.003503	0.003503
II	59	0.099662	1,089,641	0.046179	0.080741	0.004602	0.008047
III	59	0.099662	1,285,759	0.054491	0.135231	0.005431	0.013477
IV	59	0.099662	1,951,546	0.082707	0.217938	0.008243	0.02172
V	59	0.099662	1,608,364	0.068163	0.286101	0.006793	0.028513
VI	59	0.099662	1,951,546	0.082707	0.368807	0.008243	0.036756
VII	59	0.099662	2,250,460	0.095375	0.464182	0.009505	0.046261
VIII	59	0.099662	2,951,561	0.125087	0.589269	0.012466	0.058728
IX	59	0.099662	3,818,192	0.161815	0.751084	0.016127	0.074855
X	60	0.101351	5,873,413	0.248916	1	0.025228	0.101351
Total	592	1	23,596,003	1		0.100141	0.393212

Source: Author's calculations.

d_{yi} –the share of total volume of the attribute (flat value) of i-group;
d_{yi}^K –accumulated part of the total volume of the attribute (housing value) of i-group;

The total Gini ratio of the wealth distribution, expressed as the value of the housing, is calculated as follows (see formula 1):

$$G = 1 - 2 * 0,393212 + 0,100141 = 0,313717.$$

Similarly, the Gini coefficient for the distribution of wealth, expressed in value of housing in major cities, is calculated:

$$G = 1 - 2 * 0.368103 + 0.099919 = 0.363714.$$

Gini coefficient of distribution of wealth, expressed as the value of housing in the city:

$$G = 1 - 2 * 0,379817 + 0,100149 = 0,340515.$$

Gini coefficient of distribution of wealth, expressed as the value of the housing in the rural area:

$$G = 1 - 2 * 0.388152 + 0.099469 = 0.313717.$$

The Gini ratio of the wealth distribution, expressed as the value of the housing, is given in Table 3.10.

Table 3.10 Gini ratio of wealth denominated in assets

Total	In the major cities	Cities	Rural area
0.31	0.36	0.34	0.31

Source: Author's calculations.

In conclusion, it can be said that:

- The value of the housing of the Lithuanian population varies by 16.9 times. When comparing the average housing prices of I and X deciles, the difference is 7.2 times.
- Based on the preliminary assessment, and taking into account the market prices of housing, land, durable goods and other assets in 2016, and the data collected during the survey, the estimated wealth distribution is significantly higher than the income and consumption distribution values reported by official statistics. The wealth distribution coefficient (ratio of I and X deciles) is as high as 40.8 and income inequality is 11.04. Thus, survey revealed wealth inequality significantly exceeding income and consumption inequality in Lithuania.
- According to the responses of the richest respondents in the sample, 10% of the richest people in Lithuania are possessing 34.8% and 1% of the richest population in Lithuania are possessing 8.9% of all wealth of Lithuanian population. Also, when calculating the differences between adjacent decile, it can be sees that the most pronounced distances are from the tenth (X) and ninth (IX) deciles, and the first (I) and second (II) deciles reaching 2.8 times. This indicates that the value of the wealth of extreme decile groups has a significant impact on wealth inequality rates in Lithuania.
- According to the calculations of the Gini coefficient carried out to assess wealth inequality in Lithuania and separately in rural areas, towns, and cities of Lithuania, it is clear that the wealth differences of households are quite large – the Gini coefficient exceeds 0.3, i.e., the theoretical threshold of this coefficient. Therefore, after assessing the limitations of the method of determining the inequality of this wealth, it can be said that real wealth inequality in Lithuania is above income and consumption inequality reported by official statistics.

3.4 The negative impact of excessive inequality on economic growth of EU member states

The results of the empirical research on the impact of economic inequality on economic growth are presented in the following sequence:

1 the impact of increasing economic inequality on the annual economic growth rate based on the Least Square Method (LSM) and the Fixed Effects Method (FEM);

2   the impact of economic inequality on the average economic growth rate of the two-year period;
3   the impact of economic inequality on the average economic growth rate of the three-year period;
4   the impact of excessive inequality on the average economic growth rate of the four-year period; and
5   the impact of excessive inequality on the average economic growth rate of the five-year period.

The assessment of the impact of increasing economic inequality on annual economic growth of EU Member States under the *Smallest Squares Approach (SSA)*. The impact of economic inequality on annual economic growth assessed by the LSM shall examine the hypothesis (H0) that all excluded factors do not affect economic growth. The results presented in Table 3.11 show that the complex significance criterion F (p-value) is low, demonstrating the high reliability of the model, which results in the hypothesis rejection and statement that these factors are affecting economic growth. Thus, economic growth depends on the following indicators set out:

1   the initial level of GDP per capita;
2   investments (gross fixed capital formation) (% of GDP);
3   the proportion of population with tertiary education level (%);
4   inflation rate (average annual change) (%);
5   general government expenditures (% of GDP);
6   the level of government debt (% of GDP);
7   corruption perception index; and
8   volume of trade (imports, exports) (% of GDP).

The following revenue differentiation indicators have been distinguished for the research:

1   Gini coefficient of equivalized disposable income (Gini 1);
2   Gini coefficient of equivalized disposable income before social transfers, where pensions are included in social transfers (Gini 2);
3   Gini factor of equivalized disposable income before social transfers, where pensions are eliminated from social transfers (Gini 3);
4   X/I deciles' differentiation coefficient; and
5   S80/S20 income quintile ratio.

According to the calculations carried out (see Table 3.11), the indicators of economic inequality measured by the Gini coefficient of equivalized disposable income before social transfers – where pensions are included in social transfers (Gini 2) and the Gini coefficient of equivalized disposable income before social transfers, where pensions are eliminated from social transfers (Gini 3) – are statistically significant for annual economic growth.

Table 3.11 The impact of economic inequality on socio-economic progress, reflecting through economic growth indicators and by controlling other economic growth factors (dependent variable is logarithmic differentiated GDP per capita)

Variables	Gini 1	Gini 2	Gini 3	DK	S80/S20
	Coefficient estimates calculated using SSA				
	Indicators of economic inequality				
Dependent variable	Annual economic growth rate (logarithmic differentiated GDP per capita)				
Constant	0.257	−5.204*	3.725**	0.374	0.715
Change in natural logarithm of GDP per capita ($t-1$)	−0.023***	−0.024***	−0.025***	−0.025***	−0.024***
Change in the natural logarithm of investments	0.027*	0.015	0.021	0.026*	0.027*
Change in the natural logarithm of the share of population with tertiary education level	−0.014	−0.013	−0.027**	−0.011	−0.013
Inflation rate	0.002	0.003	0.003	0.002	0.002
Change in the natural logarithm of general government expenditure	−0.044**	−0.055***	−0.014	−0.047**	−0.043**
Change in the natural logarithm of government debt	0.009*	0.007	0.005	0.011**	0.011**
Change in the natural logarithm of the corruption perception index	0.049**	0.037*	0.047**	0.055**	0.054**
Change in natural logarithm of trade (imports, exports)	0.025***	0.025***	0.027***	0.027***	0.027***
The natural logarithm of economic inequality	−0.174	2.712*	−2.194**	−0.121	−0.259
The quadratic natural logarithm of economic inequality	0.033	−0.341*	0.325***	0.011	0.023
N (number of observations)	196	196	196	196	196
Adjusted R^2	0.360	0.371	0.420	0.360	0.364
F criteria (*p*-value)	<0.001	<0.001	<0.001	<0.001	<0.001
The value of test p to check for autocorrelation	0.014	0.036	0.215	0.012	0.000
Value of test p to check heteroskedasticity	0.057	0.092	0.002	0.069	0.071
Breaking point	–	53.46	29.35	–	–
Actual range of the economic inequality indicator	from 20.9 to 40.8	from 37.2 to 61.6	from 24.3 to 46.8	from 4.49 to 18.07	from 3.03 to 8.32

Source: Authors' calculations.

Note: Gini 1 – Gini coefficient of equivalized disposable income; Gini 2 – Gini coefficient of equivalized disposable income before social transfers (pensions included in social transfers); Gini 3 – Gini coefficient of equivalized disposable income before social transfers (pensions excluded from social transfers); DK – X/I deciles' differentiation coefficient; S80/S20 – income quintile ratio.

* 90% of significance level.

** 95% of significance level.

*** 99% of significance level.

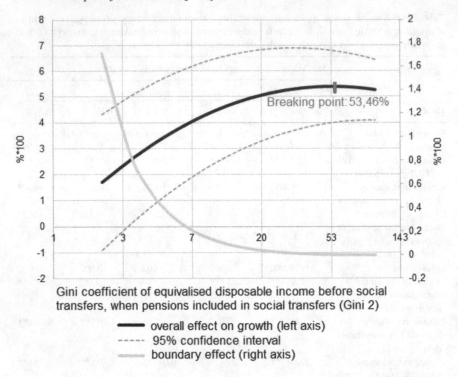

Figure 3.18 Relative slope of the impact of economic inequality (according to Gini 2) on socio-economic progress, expressed by economic growth indicators.

Source: Authors' calculations.

The increase of economic inequality assessed by Gini 2 coefficient (Gini coefficient of equivalized disposable income before social transfers where pensions are included in social transfers) until the identified breaking point (53.46%) has a positive marginal effect on economic growth, i.e., greater economic inequality provides for faster economic growth, but when the breaking point is reached, the marginal effect of economic inequality becomes negative and the further increase in economic inequality is linked to slowing economic growth (see Figure 3.18).

Taking into account fact that the Gini 2 indicator highlights the real level of economic inequality, before the state's social assistance functions were implemented, it is necessary to stress that excessive inequality and higher levels than breaking point were recorded in Sweden in 2018 (57.1%), Greece (57%), Portugal (56.5%), Bulgaria (54.8%), Romania (54.6%), Germany (56.4%), and the United Kingdom (53.7%). In this way, it can be argued that it is the economic inequality of the mentioned countries have a negative impact on their economic growth. In Lithuania, the Gini coefficient of equivalized disposable income before social transfers, when pensions are included in social transfers, in 2018 was 51.1% and is very close to the threshold or breaking point where the impact of increasing economic inequality has negative impact on economic growth.

The increase of economic inequality expressed by Gini 3 coefficient (Gini coefficient of equivalized disposable income before social transfers, where pensions are eliminated from social transfers) has a positive marginal effect on economic growth until the fixed breaking point (29.35%). After this threshold, the marginal effect of economic inequality becomes negative and the further increase in economic inequality is linked to slowing economic growth (see Figure 3.19). The different directions of economic inequality measured by Gini 2 and Gini 3 depend on the state's social security and redistributive policies, and more specifically on social pensions' levels.

The Chow test was applied to examine the hypothesis (H0) that the impact of economic inequality on economic growth in different groups of countries – countries with higher living standards GDP per capita measured in purchasing power standards (PPS) are more than €24750 and countries with lower living standards (GDP per capita at PPS is less than

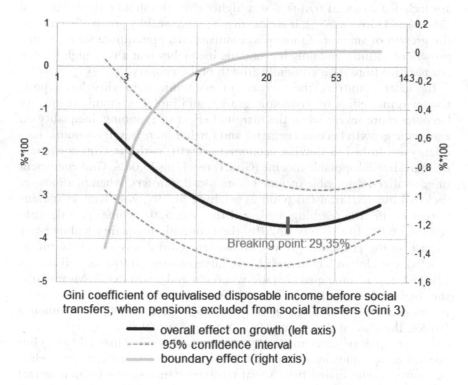

Gini coefficient of equivalised disposable income before social transfers, when pensions excluded from social transfers (Gini 3)

———— overall effect on growth (left axis)
- - - - 95% confidence interval
———— boundary effect (right axis)

Figure 3.19 Relative slope of the impact of economic inequality (according to Gini 3) on socio-economic progress, expressed by economic growth indicators.

Source: Authors' calculations.

€24750) are the same in both groups. Despite the fact that statistically significant indicators of economic inequality did not emerge in the present case, the Chow test's complex criterion of significance of the indicators F (*p*-value) is very low, demonstrating the high reliability of the calculations, which results in the hypothesis being rejected. Therefore, the impact of economic inequality on economic growth varies between the identified groups of countries.

In less wealthy groups of countries, economic inequality, measured by the Gini coefficient of equivalized disposable income before social transfers, when pensions are included in social transfers (Gini 2) and Gini coefficient of equivalized disposable income before social transfers, where pensions are eliminated from social transfers (Gini 3), the marginal effect of inequality is positive, i.e., increasing economic inequality stimulate economic growth and only at relatively high levels of economic inequality – a breaking point at 55.84% and 47.84%, respectively, for Gini 2 and Gini 3, the marginal effect becomes negative and the increase of economic inequality is starting to have a negative impact on economic growth. In 2018, Gini 2 (Gini coefficient of equivalized disposable income before social transfers, when pensions are included in social transfers) was higher than threshold only in Portugal (56.5%) and Greece (57%). It is clear that in a less wealthy group of countries, the growth of amount of money accumulated is a prerequisite for economic growth of country and only if economic inequality is at a very high level; it has negative impact on economic growth of the country.

In richer countries, the increase in economic inequality has a positive marginal effect on economic growth and just if economic inequality becomes more severe when the marginal effect of economic inequality on economic growth becomes negative and further increases in economic inequality are linked to slowing economic growth, with the Gini coefficient of equivalized disposable income (Gini 1) reaching 35.06%, Gini coefficient of equivalized disposable income before social transfers, when pensions are included in social transfers (Gini 2) reaching 51.74%, X/I deciles differentiation coefficient reaching 8.96, and the S80/S20 income quintile ratio reaching 6.00. In this way in 2018, the economic inequality had negative impact on the economic growth in Sweden (Gini 2 was 57.1%), Germany (56.4%), the United Kingdom (53.7%), Luxembourg (51.9%), as well as Italy (the S80/S20 income quartile ratio was 6.09) and Spain (6.03). Meanwhile, the higher X/I coefficient of deciles differentiation than the above-mentioned level was fixed in Italy (12.55), Spain (11.00), Luxembourg (10.33), the United Kingdom (9.96), and Germany (9.88).

In a group of richer countries, by measuring economic inequality by Gini coefficient of equivalized disposable income before social transfers, when pensions are eliminated from social transfers (Gini 3), the latter marginal effect has a negative effect on economic growth and only after reaching the breaking point of 32.37% does the marginal effect of economic inequality become positive on economic growth.

For assessment of the reliability of the LSM model, the tests were carried out. First, the Wooldridge test for autocorrelation, where the hypothesis (H0) is tested, that for the model is autocorrelation is characteristic, but the average *p*-value obtained is 0.055 (see Table 3.12), suggesting that there is no autocorrelation in model. Secondly, the White test was applied to determine heteroskedism when testing a hypothesis (H0) that the model was characterized by heteroskedism, but that the mean *p*-value of the test was 0.058 (see Table 3.12), so it should be concluded that there was no heteroskedism in the model. Third, the diagnostics of the panel data was carried based on Hausman test and the Fixed Effects Method (FEM) is recommended in the next stage of the research.

The impact of inequality on annual economic growth under the FEM. The assessment of the impact of economic inequality on annual economic growth by the FEM shall examine the hypothesis (H0) that all excluded factors do not affect economic growth. Based on the results presented in Table 3.13,

Table 3.12 The impact of economic inequality on socio-economic progress by group of EU countries, reflecting through economic growth indicators and controlling other economic growth factors (dependent variable is logarithmic differentiated GDP per capita)

	Coefficient estimates calculated using SSA				
	Indicators of economic inequality				
Variables	Gini 1	Gini 2	Gini 3	DK	S80/S20
Dependent variable	Annual economic growth rate (logarithmic differentiated GDP per capita)				
	Less wealthy countries				
Constant	−0.888	−6.796	−2.623	−0.601	−1.239
Change in natural logarithm of GDP per capita ($t-1$)	−0.025	−0.016	−0.015	−0.035*	−0.028
Change in the natural logarithm of investments	0.027	0.005	0.021	0.020	0.027
Change in the natural logarithm of the share of population with tertiary education level	−0.038*	−0.038	−0.042***	−0.035	−0.036
Inflation rate	0.000	−0.001	−0.002	0.000	0.000
Change in the natural logarithm of general government expenditure	0.002	−0.034	−0.018	0.004	0.008
Change in the natural logarithm of government debt	0.004	−0.003	0.000	0.007	0.007
Change in the natural logarithm of the corruption perception index	0.057	0.051	0.044	0.083**	0.070*
Change in natural logarithm of trade (imports, exports)	0.041***	0.036***	0.038***	0.048***	0.049***
The natural logarithm of economic inequality	0.364	3.430	1.363	0.084	0.263

(Continued)

Table 3.12 The impact of economic inequality on socio-economic progress by group of EU countries, reflecting through economic growth indicators and controlling other economic growth factors (dependent variable is logarithmic differentiated GDP per capita) *(Continued)*

	Coefficient estimates calculated using SSA				
	Indicators of economic inequality				
Variables	Gini 1	Gini 2	Gini 3	DK	S80/S20
The quadratic natural logarithm of economic inequality	−0.039	−0.426	−0.176	−0.003	−0.016
Breaking point	−	55.84	47.84	−	−
	The difference between rich and less rich countries				
The pseudo-variable of richer countries	0.207	−19.211	10.531	0.034	0.553
Change in natural logarithm of GDP per capita ($t-1$)	−0.014	−0.047	−0.024	−0.003	−0.010
Change in the natural logarithm of investments	−0.008	0.007	0.002	−0.003	−0.009
Change in the natural logarithm of the share of population with tertiary education level	0.049	0.030	0.027	0.049	0.048
Inflation rate	0.001	0.005	0.005	0.001	0.002
Change in the natural logarithm of general government expenditure	−0.056	−0.031	0.018	−0.061	−0.064
Change in the natural logarithm of government debt	0.002	0.007	−0.002	−0.001	−0.001
Change in the natural logarithm of the corruption perception index	−0.036	−0.003	−0.027	−0.072	−0.053
Change in natural logarithm of trade (imports, exports)	−0.015	0.000	−0.010	−0.024	−0.024
The natural logarithm of economic inequality	0.200	10.034	−5.790	0.188	0.055
The quadratic natural logarithm of economic inequality	−0.041	−1.280	0.813	−0.017	−0.009
Breaking point	35.06	51.74	32.37	8.96	6.00
Actual range of the economic inequality indicator	from 20.9 to 40.8	from 37.2 to 61.6	from 24.3 to 46.8	from 4.49 to 18.07	from 3.03 to 8.32
N (number of observations)	168	168	168	168	168
Adjusted R^2	0.393	0.421	0.456	0.396	0.399
F criteria (p-value)	<0.001	<0.001	<0.001	<0.001	<0.001

Source: Authors' calculations.

Note: Gini 1 – Gini coefficient of equivalized disposable income; Gini 2 – Gini coefficient of equivalized disposable income before social transfers (pensions included in social transfers); Gini 3 – Gini coefficient of equivalized disposable income before social transfers (pensions excluded from social transfers); DK – X/I deciles' differentiation coefficient; S80/S20 – income quintile ratio.

* 90% of significance level.

** 95% of significance level.

*** 99% of significance level.

Table 3.13 The impact of economic inequality on socio-economic progress, reflecting through economic growth indicators and controlling other economic growth factors (the dependent variable is logarithmic differentiated GDP per capita)

	Coefficient estimates calculated using FEM				
	Indicators of economic inequality				
Variables	Gini 1	Gini 2	Gini 3	DK	S80/S20
Dependent variable	Annual economic growth rate (logarithmic differentiated GDP per capita)				
Constant	1.842	−1.964	2.254	2.034	3.695
Change in natural logarithm of GDP per capita ($t - 1$)	−0.077	−0.067	−0.069	−0.079	−0.084★
Change in the natural logarithm of investments	−0.053★	−0.055★	−0.052★	−0.055★	−0.055★
Change in the natural logarithm of the share of population with tertiary education level	0.011	0.018	0.013	0.006	0.008
Inflation rate	0.003	0.004	0.003	0.003	0.003
Change in the natural logarithm of general government expenditure	0.042	0.046	0.037	0.049	0.045
Change in the natural logarithm of government debt	0.015	0.017	0.014	0.014	0.011
Change in the natural logarithm of the corruption perception index	−0.011	−0.003	−0.007	−0.013	−0.012
Change in natural logarithm of trade (imports, exports)	0.134★★	0.141★★★	0.145★★★	0.130★★	0.132★★
The natural logarithm of economic inequality	−0.980	0.949	−1.273	−0.525	−1.086
The quadratic natural logarithm of economic inequality	0.139	−0.120	0.179	0.037	0.085
N (number of observations)	196	196	196	196	196
Adjusted R^2	0.335	0.333	0.335	0.339	0.338
F criteria (p-value)	<0.001	<0.001	<0.001	<0.001	<0.001
Breaking point	33.54	52.93	35.00	12.21	6.00
Actual range of the economic inequality indicator	from 20.9 to 40.8	from 37.2 to 61.6	from 24.3 to 46.8	from 4.49 to 18.07	from 3.03 to 8.32

Source: Authors' calculations.

Note: Gini 1 – Gini coefficient of equivalized disposable income; Gini 2 – Gini coefficient of equivalized disposable income before social transfers (pensions included in social transfers); Gini 3 – Gini coefficient of equivalized disposable income before social transfers (pensions excluded from social transfers); DK – X/I deciles' differentiation coefficient; S80/S20 – income quintile ratio.

★ 90% of significance level.

★★ 95% of significance level.

★★★ 99% of significance level.

the complex significance criterion F (p-value) of the indicators is low that demonstrates the high reliability of the calculations and results obtained, which results in the hypothesis being rejected. Therefore, as in the case of the LSM, the economic inequality has positive impact on economic growth.

Assessing the economic inequality on the basis of the Gini coefficient of equivalized disposal income (Gini 1) and Gini coefficient of equivalized disposable income before social transfers, where pensions are eliminated from social transfers (Gini 3), the X/I deciles coefficient of differentiation and the 80/S20 income quintile ratio, the economic inequality has a negative marginal effect on economic growth and only when it reaches a breaking point of 33.54%, 35%, 12.21, and 6.00 respectively, the marginal effect becomes a positive effect on economic growth measured by logarithmic differentiated GDP per capita. Meanwhile, when measuring economic inequality by Gini coefficient of equivalized on disposable income before social transfers, where pensions are included in social transfers (Gini 2), the results are opposite. The identified breaking point (52.93%) shows that the increase in economic inequality has a positive marginal effect on economic growth (greater economic inequality is linked to faster economic growth), but when this breaking point is reached, the marginal effect of economic inequality becomes negative and the further increase in economic inequality is linked to slowing economic growth (see Table 3.14). It should be emphasized that the Gini 2 indicator highlights the real level of economic inequality before the state implements the functions of social assistance. Moreover, as in the case of the LSM, there is the same tendency that the directions of economic inequality impact on economic growth depends on the state's policies for ensuring social security and redistribution of money as well upon pensions' levels.

The EU-28 countries are divided into countries with higher living standards (where GDP per capita in PPS is more than €24750) and countries with lower living standards (where GDP per capita in PPS is less than €24750). An analysis of the impact of economic inequality on economic growth in these groups was assessed and compared. The statistically significant impact of economic inequality on economic growth is highlighted by using Gini 3 coefficient of equivalized disposable income before social transfers, when pensions are eliminated from social transfers in the group of richer countries (see Table 3.14).

According to the results obtained (see Table 3.14), in richer countries, the increase in economic inequality has a positive marginal effect on economic growth, measured by a logarithmic differentiated GDP per capita until the of reaching a breaking point at the 34.03% according to Gini coefficient of equivalized disposable income before social transfers, when pensions are eliminated from social transfers (Gini 3). After reaching this point, the marginal effect of economic inequality measured by Gini 3 becomes negative and the further increase in economic inequality is linked to slowing economic growth. In this way, in 2018 the economic inequality has a negative impact on the economic growth in the vast majority of the wealthier group:

Table 3.14 The impact of economic inequality on socio-economic progress by group of countries, reflecting through economic growth indicators and controlling other growth factors (dependent variable is logarithmic differentiated GDP per capita)

Variables	Coefficient estimates calculated using FEM				
	Indicators of economic inequality				
	Gini 1	*Gini 2*	*Gini 3*	*DK*	*S80/S20*
Dependent variable	Annual economic growth rate (logarithmic differentiated GDP per capita)				
	Less wealthy countries				
Constant	1.262	−1.207	0.881	1.530	2.111
Change in natural logarithm of GDP per capita ($t-1$)	−0.067	−0.063	−0.102*	−0.079	−0.081
Change in the natural logarithm of investments	−0.024	−0.026	−0.030	−0.026	−0.029
Change in the natural logarithm of the share of population with tertiary education level	0.006	0.014	0.033	0.014	0.017
Inflation rate	0.002	0.003	0.002	0.002	0.002
Change in the natural logarithm of general government expenditure	0.016	0.018	0.029	0.021	0.019
Change in the natural logarithm of government debt	−0.006	−0.008	−0.017	−0.008	−0.009
Change in the natural logarithm of the corruption perception index	−0.049	−0.045	−0.055	−0.054	−0.055
Change in natural logarithm of trade (imports, exports)	0.133**	0.121**	0.116**	0.125**	0.131**
The natural logarithm of economic inequality	−0.529	0.725	−0.139	−0.293	−0.515
The quadratic natural logarithm of economic inequality	0.080	−0.082	0.030	0.021	0.042
Breaking point	27.67	−	−	9.63	4.75
	The difference between rich and less rich countries				
The pseudo-variable of richer countries	−1.177	−20.440	−10.902	0.458	−3.915
Change in natural logarithm of GDP per capita ($t-1$)	−0.275***	−0.276***	−0.296***	−0.260***	−0.262***
Change in the natural logarithm of investments	−0.037	−0.034	−0.024	−0.036	−0.034
Change in the natural logarithm of the share of population with tertiary education level	−0.019	−0.022	−0.041	−0.036	−0.037
Inflation rate	0.003	0.003	0.002	0.002	0.003
Change in the natural logarithm of general government expenditure	−0.206*	−0.160	−0.115	−0.208*	−0.210*
Change in the natural logarithm of government debt	0.019	0.014	0.020	0.020	0.019

(Continued)

Table 3.14 The impact of economic inequality on socio-economic progress by group of countries, reflecting through economic growth indicators and controlling other growth factors (dependent variable is logarithmic differentiated GDP per capita)   *(Continued)*

	Coefficient estimates calculated using FEM				
	Indicators of economic inequality				
Variables	Gini 1	Gini 2	Gini 3	DK	S80/S20
Change in the natural logarithm of the corruption perception index	0.092	0.078	0.101	0.106	0.110
Change in natural logarithm of trade (imports, exports)	−0.042	−0.029	−0.048*	−0.051	−0.056
The natural logarithm of economic inequality	2.792	12.314	8.155*	0.887	2.407
The quadratic natural logarithm of economic inequality	−0.422	−1.597	−1.167*	−0.068	−0.199
Breaking point	27.35	48.57	34.03	5.66	4.16
Actual range of the economic inequality indicator	from 20.9 to 40.8	from 37.2 to 61.6	from 24.3 to 46.8	from 4.49 to 18.07	from 3.03 to 8.32
N (number of observations)	168	168	168	168	168
Adjusted R^2	0.428	0.440	0.441	0.432	0.431
F criteria (p-value)	<0.001	<0.001	<0.001	<0.001	<0.001

Source: Authors' calculations.

Note: Gini 1 – Gini coefficient of equivalized disposable income; Gini 2 – Gini coefficient of equivalized disposable income before social transfers (pensions included in social transfers); Gini 3 – Gini coefficient of equivalized disposable income before social transfers (pensions excluded from social transfers); DK – X/I deciles' differentiation coefficient; S80/S20 – income quintile ratio.

* 90% of significance level.

** 95% of significance level.

*** 99% of significance level.

as Gini 3 was 40.4% in the United Kingdom, 39.3% – in Ireland, 38.1% – in Luxembourg, 37% – in Spain 36.6% in Germany, 36% in Denmark, 35.7% – in Sweden 35.7% and Italy, 34.9% – in France, 34.4% – in Finland, and 34.1% in Estonia.

The impact of economic inequality on the average rate of two-year economic growth. The assessment of the impact of economic inequality on the average two-year growth rate highlights the statistical significance of the X/I deciles' coefficient of differentiation and the 80/S20 income quintile ratio. According to the calculations and results obtained (see Table 3.15), economic inequality has a negative impact on economic growth and only when it reaches a breaking point of 11.19 and 5.94, respectively, is the positive effect on economic growth can be observed. The Chow test for the latter model (see Table 3.16) shows the statistical significance of the X/I deciles' differentiation rate and the S80/S20 income quintile ratio, where economic inequality adversely affects economic growth and, at that

Table 3.15 The impact of economic inequality on socio-economic progress, reflecting economic growth indicators and controlling other growth factors (dependent variable is average two-year economic growth rate)

	Coefficient estimates calculated using FEM				
	Indicators of economic inequality				
Variables	Gini 1	Gini 2	Gini 3	DK	S80/S20
Dependent variable	Average two-year economic growth rate				
Constant	5.126*	−1.788	3.594	4.628**	8.122**
Change in natural logarithm of GDP per capita ($t − 2$)	−0.131***	−0.113***	−0.113***	−0.131***	−0.139***
Change in the natural logarithm of investments	−0.067***	−0.070***	−0.066***	−0.068***	−0.068***
Change in the natural logarithm of the share of population with tertiary education level	0.033	0.038	0.036	0.025	0.026
Inflation rate	−0.002	−0.001	−0.001	−0.002	−0.002
Change in the natural logarithm of general government expenditure	0.020	0.024	0.017	0.034	0.027
Change in the natural logarithm of government debt	0.012	0.020	0.015	0.012	0.007
Change in the natural logarithm of the corruption perception index	−0.027	−0.015	−0.018	−0.022	−0.025
Change in natural logarithm of trade (imports, exports)	0.081*	0.089**	0.096**	0.082**	0.079*
The natural logarithm of economic inequality	−2.327	1.287	−1.595	−1.032**	−2.183**
The quadratic natural logarithm of economic inequality	0.333	−0.164	0.224	0.074**	0.171**
N (number of observations)	168	168	168	168	168
Adjusted R^2	0.451	0.440	0.443	0.462	0.463
F criteria (p-value)	<0.001	<0.001	<0.001	<0.001	<0.001
Breaking point	32.94	50.67	35.06	11.19	5.94
Actual range of the economic inequality indicator	from 20.9 to 40.8	from 37.2 to 61.6	from 24.3 to 46.8	from 4.49 to 18.07	from 3.03 to 8.32

Source: Authors' calculations.

Note: Gini 1 – Gini coefficient of equivalized disposable income; Gini 2 – Gini coefficient of equivalized disposable income before social transfers (pensions included in social transfers); Gini 3 – Gini coefficient of equivalized disposable income before social transfers (pensions excluded from social transfers); DK – X/I deciles' differentiation coefficient; S80/S20 – income quintile ratio.

* 90% of significance level.

** 95% of significance level.

***99% of significance level.

breaking point, the economic inequality is starting to have a positive impact on economic growth of less wealthy EU countries. It can be assumed that the richest members of society invest and contribute to the country's well-being as their income constantly increases.

According to the calculations carried out (see Table 3.16), the statistical significance of Gini coefficient of equivalized disposable income before social transfers, pensions eliminated from social transfers (Gini 3), is revealed in the group of richer countries. In the group of richer countries, the increase in

Table 3.16 Impact of economic inequality on socio-economic progress by group of countries, reflecting through economic growth indicators and controlling other economic growth factors (the dependent variable is average two-year growth rate)

| | *Coefficient estimates calculated using FEM* | | | | |
| | *Indicators of economic inequality* | | | | |
Variables	*Gini 1*	*Gini 2*	*Gini 3*	*DK*	*S80/S20*
Dependent variable	Average two-year economic growth rate				
	Less wealthy countries				
Constant	3.859	−3.307	1.076	4.951**	7.239**
Change in natural logarithm of GDP per capita $(t-2)$	−0.097**	−0.076*	−0.121***	0.104**	−0.113**
Change in the natural logarithm of investments	−0.046	−0.042	−0.048	−0.040	−0.051
Change in the natural logarithm of the share of population with tertiary education level	0.054	0.059	0.043	0.058	0.063
Inflation rate	−0.002	−0.001	−0.002	−0.003	−0.002
Change in the natural logarithm of general government expenditure	0.003	0.006	0.012	0.013	0.009
Change in the natural logarithm of government debt	−0.008	0.000	0.001	−0.006	−0.007
Change in the natural logarithm of the corruption perception index	−0.047	−0.039	−0.047	−0.053	−0.055
Change in natural logarithm of trade (imports, exports)	0.089**	0.087*	0.058	0.078**	0.080**
The natural logarithm of economic inequality	−1.737	1.989	0.051	−1.138**	−1.936*
The quadratic natural logarithm of economic inequality	0.253	−0.254	0.000	0.081**	0.151*
Breaking point	31.05	50.46	*0.00*	11.62	6.00

(Continued)

Table 3.16 Impact of economic inequality on socio-economic progress by group of countries, reflecting through economic growth indicators and controlling other economic growth factors (the dependent variable is average two-year growth rate)   *(Continued)*

Variables	Coefficient estimates calculated using FEM				
	Indicators of economic inequality				
	Gini 1	Gini 2	Gini 3	DK	S80/S20
	The difference between rich and less rich countries				
The pseudo-variable of richer countries	−7.546	−9.012	−23.888***	0.445	−8.386
Change in natural logarithm of GDP per capita $(t − 2)$	−0.294***	−0.326***	−0.399***	−0.307***	−0.287***
Change in the natural logarithm of investments	−0.023	−0.029	−0.056	−0.023	−0.015
Change in the natural logarithm of the share of population with tertiary education level	−0.055	−0.045	−0.030	−0.064	−0.070
Inflation rate	0.003	0.003	0.002	0.002	0.002
Change in the natural logarithm of general government expenditure	−0.278***	−0.279***	−0.155	−0.277***	−0.267**
Change in the natural logarithm of government debt	0.059	0.054	0.004	0.048	0.044
Change in the natural logarithm of the corruption perception index	0.055	0.000	0.060	0.048	0.070
Change in natural logarithm of trade (imports, exports)	−0.058**	−0.024	−0.043**	−0.038	−0.060*
The natural logarithm of economic inequality	6.960	6.889	16.196***	1.103	4.024
The quadratic natural logarithm of economic inequality	−1.045	−0.884	−2.285***	−0.082	−0.328
Breaking point	26.96	49.50	34.98	–	3.71
Actual range of the economic inequality indicator	from 20.9 to 40.8	from 37.2 to 61.6	from 24.3 to 46.8	from 4.49 to 18.07	from 3.03 to 8.32
N (number of observations)	168.000	168.000	168.000	168.000	168.000
Adjusted R^2	0.614	0.609	0.648	0.620	0.620
F criteria (p-value)	<0.001	<0.001	<0.001	<0.001	<0.001

Source: Authors' calculations.

Note: Gini 1 – Gini coefficient of equivalized disposable income; Gini 2 – Gini coefficient of equivalized disposable income before social transfers (pensions included in social transfers); Gini 3 – Gini coefficient of equivalized disposable income before social transfers (pensions excluded from social transfers); DK – X/I deciles' differentiation coefficient; S80/S20 – income quintile ratio.

* 90% of significance level.

** 95% of significance level.

*** 99% of significance level.

economic inequality has a positive marginal effect on average two-year economic growth; at a breaking point of 34.98% in Gini 3, the marginal effect of economic inequality on economic growth becomes negative and further increases in economic inequality are linked to slowing economic growth (see Table 3.16). Thus, according to 2018 data, economic inequality has a negative impact on the economic growth in the United Kingdom, as Gini 3 was 40.4%, Ireland (39.3%), Luxembourg (38.1%), Spain (37%), Germany (36.6%), Denmark (36%), Sweden (35.7%), and Italy (35.7%).

The impact of economic inequality on the average economic growth rate of the three-year period. The analysis of the impact of economic inequality on the average economic growth rate of the three-year reveals the statistical significance of the Gini coefficient of equivalized disposable income (Gini 1), the X/I deciles' differentiation coefficient, and the S80/S20 income quintile ratio. According to the calculations and results obtained (see Table 3.17), the marginal effect of economic inequality on economic growth is negative and only at a breaking point of 29.18%, 9.01 and 4.99, respectively, the marginal effect of economic inequality on economic growth becomes a positive. The Chow test (see Table 3.18) shows that the statistical significance for the Gini 1 coefficient, X/I deciles' differentiation coefficient, and the S80/S20 income quintile ratio, where economic inequality has a negative marginal effect on economic growth and, when the above-mentioned breaking point is reached, the marginal effect of economic inequality becomes positive for economic growth, which is typical for lower living standard EU countries. Similar results were obtained by analyzing the impact of economic inequality on the average two-year growth rate.

In the group of richer countries, Gini coefficient of equivalized disposable income (Gini 1), Gini coefficient of equivalized disposable income before social transfers, when pensions are included in social transfers (Gini 2) and Gini coefficient of equivalized disposable income before social transfers, when pensions are eliminated from social transfers (Gini 3) are statistically significant. In the wealthier group, the marginal effect of economic inequality on average three-year economic growth is positive and, in just then Gini 1 reaches 29.63%, Gini 2 – 53.06%, and Gini 3 – 36.45%, the marginal effect of economic inequality on economic growth becomes negative and the further increases in economic inequality is linked to slowing economic growth (see Table 3.18). According to 2018 statistics, excessive economic inequality can be observed in the following EU countries – Luxembourg (where Gini 1 is 33.2%, Gini 2 is 51.9%, and Gini 3 is 38.1%), Germany (Gini 1 – 31.1%, Gini 2 – 56.4%, Gini 3 – 36.6%), the United Kingdom (Gini 1 – 33.5%, Gini 2 – 53.7%, Gini 3 – 40.4%), and Spain (Gini 1 – 33.2%; Gini 3 – 37%), Ireland (Gini 3 – 39.3%), Sweden (Gini 2 – 57.1%), Italy (Gini 1 – 33.4%), and Estonia (Gini 1 – 30.6%).

The impact of excessive inequality on the average growth rate of the four-year economy. Despite the fact that the assessment of the impact of economic inequality on the average economic growth rate for the four-year economy

Table 3.17 The impact of economic inequality on socio-economic progress, reflecting economic growth indicators and controlling other growth factors (dependent variable – average three-year economic growth rate)

| Variables | Coefficient estimates calculated using FEM | | | | |
| | Indicators of economic inequality | | | | |
	Gini 1	Gini 2	Gini 3	DK	S80/S20
Dependent variable	Average three-year economic growth rate				
Constant	4.956**	0.462	3.137	3.630***	6.094**
Change in natural logarithm of GDP per capita $(t-3)$	−0.162***	−0.146***	−0.141***	−0.164***	−0.169***
Change in the natural logarithm of investments	−0.048***	−0.047***	−0.047***	−0.045***	−0.046***
Change in the natural logarithm of the share of population with tertiary education level	0.027	0.025	0.028	0.022	0.023
Inflation rate	−0.005***	−0.006***	−0.005***	−0.006***	−0.005***
Change in the natural logarithm of general government expenditure	0.033	0.039	0.033	0.040	0.037
Change in the natural logarithm of government debt	0.000	0.008	0.001	0.001	−0.003
Change in the natural logarithm of the corruption perception index	−0.015	−0.008	−0.007	−0.011	−0.015
Change in natural logarithm of trade (imports, exports)	0.042	0.034	0.050	0.038	0.042
The natural logarithm of economic inequality	−2.067*	0.407	−1.159	−0.637*	−1.467*
The quadratic natural logarithm of economic inequality	0.306*	−0.051	0.170	0.047*	0.118*
N (number of observations)	140	140	140	140	140
Adjusted R^2	0.628	0.612	0.620	0.628	0.634
F criteria (p-value)	<0.001	<0.001	<0.001	<0.001	<0.001
Breaking point	29.18	52.57	30.13	9.01	4.99
Actual range of the economic inequality indicator	from 20.9 to 40.8	from 37.2 to 61.6	from 24.3 to 46.8	from 4.49 to 18.07	from 3.03 to 8.32

Source: Authors' calculations.

Note: Gini 1 – Gini coefficient of equivalized disposable income; Gini 2 – Gini coefficient of equivalized disposable income before social transfers (pensions included in social transfers), Gini 3 – Gini coefficient of equivalized disposable income before social transfers (pensions excluded from social transfers); DK – X/I deciles' differentiation coefficient; S80/S20 – income quintile ratio

* 90% of significance level.

** 95% of significance level.

*** 99% of significance level.

Table 3.18 The impact of economic inequality on socio-economic progress by group of countries, reflecting through economic growth indicators and controlling other economic growth factors (the dependent variable is average three-year economic growth rate)

Variables	Coefficient estimates calculated using FEM				
	Indicators of economic inequality				
	Gini 1	Gini 2	Gini 3	DK	S80/S20
Dependent variable	Average three-year economic growth rate				
	Less wealthy countries				
Constant	3.473	−2.182	0.959	4.266★★★	6.227★★
Change in natural logarithm of GDP per capita $(t − 3)$	−0.039	−0.100★★★	−0.102★★★	−0.129★★★	−0.129★★★
Change in the natural logarithm of investments	−0.046★★	−0.029	−0.039★★	−0.035★	−0.041★★
Change in the natural logarithm of the share of population with tertiary education level	0.123★★★	0.041	0.049	0.031	0.039
Inflation rate	−0.004★★	−0.004★★	−0.003★	−0.005★★★	−0.004★★★
Change in the natural logarithm of general government expenditure	−0.015	0.013	0.028	0.021	0.018
Change in the natural logarithm of government debt	0.002	0.004	0.001	−0.004	−0.006
Change in the natural logarithm of the corruption perception index	−0.004	−0.004	−0.004	−0.011	−0.017
Change in natural logarithm of trade (imports, exports)	0.076★★	0.053★★	0.046★	0.039	0.043
The natural logarithm of economic inequality	−2.085★	1.544	−0.158	−0.881★★	−1.583★★
The quadratic natural logarithm of economic inequality	0.306★	−0.203	0.031	0.063★★	0.125★★
Breaking point	30.31	45.17	−	10.74	5.63
	The difference between rich and less rich countries				
The pseudo-variable of richer countries	−11.709	−16.399	−27.332★★★	0.779	0.100
Change in natural logarithm of GDP per capita $(t − 3)$	−0.110★	−0.118★★	−0.226★★★	−0.119★★	−0.132★★
Change in the natural logarithm of investments	0.027	0.019	−0.007	0.021	0.030
Change in the natural logarithm of the share of population with tertiary education level	−0.084★	−0.028	−0.037	−0.012	−0.018
Inflation rate	−0.001	0.004★★	0.000	0.004★	0.003

(Continued)

Table 3.18 The impact of economic inequality on socio-economic progress by group of countries, reflecting through economic growth indicators and controlling other economic growth factors (the dependent variable is average three-year economic growth rate) *(Continued)*

Variables	Coefficient estimates calculated using FEM				
	Indicators of economic inequality				
	Gini 1	Gini 2	Gini 3	DK	S80/S20
Change in the natural logarithm of general government expenditure	0.049	0.015	0.116	−0.005	−0.015
Change in the natural logarithm of government debt	−0.001	−0.005	−0.044★	−0.008	−0.005
Change in the natural logarithm of the corruption perception index	−0.116	−0.141★★	−0.095	−0.159★★	−0.141★
Change in natural logarithm of trade (imports, exports)	−0.050★★	0.006	−0.024★	0.004	0.000
The natural logarithm of economic inequality	7.983★	9.063★	16.754★★★	0.274	0.531
The quadratic natural logarithm of economic inequality	−1.176★	−1.133★	−2.339★★★	−0.018	−0.038
Breaking point	29.63	53.06	36.45	7.94	4.29
Actual range of the economic inequality indicator	from 20.9 to 40.8	from 37.2 to 61.6	from 24.3 to 46.8	from 4.49 to 18.07	from 3.03 to 8.32
N (number of observations)	140	140	140	140	140
Adjusted R^2	0.708	0.787	0.838	0.788	0.787
F criteria (p-value)	<0.001	<0.001	<0.001	<0.001	<0.001

Source: Authors' calculations.

Note: Gini 1 – Gini coefficient of equivalized disposable income; Gini 2 – Gini coefficient of equivalized disposable income before social transfers (pensions included in social transfers) Gini coefficient of equivalized disposable income before social transfers (pensions included in social transfers); Gini 3 – Gini coefficient of equivalized disposable income before social transfers (pensions excluded from social transfers); DK – X/I deciles' differentiation coefficient; S80/S20 – income quintile ratio

★ 90% of significance level.

★★ 95% of significance level.

★★★ 99% of significance level.

did not reveal a statistically significant indicator of economic inequality. The low value of the criterion of complex significance F (p-value) indicates a high reliability of calculations and results, and it is therefore considered that the identified factors in the complex have an impact on economic growth (see Table 3.19). The impact of Gini coefficient of equivalized disposable income before social transfers, when pensions are included in social transfers (Gini 2), is positive on the four-year economic growth and from

Table 3.19 The impact of economic inequality on socio-economic progress, reflecting economic growth indicators, controlling other growth factors (dependent variable – average four-year economic growth rate)

Variables	Coefficient estimates calculated using FEM				
	Indicators of economic inequality				
	Gini 1	Gini 2	Gini 3	DK	S80/S20
Dependent variable	Average four-year economic growth rate				
Constant	3.161**	−1.207	2.609	2.261**	3.640**
Change in natural logarithm of GDP per capita ($t − 4$)	−0.143***	−0.131***	−0.125***	−0.144***	−0.149***
Change in the natural logarithm of investments	−0.024*	−0.026*	−0.027*	−0.021	−0.021
Change in the natural logarithm of the share of population with tertiary education level	0.013	0.008	0.012	0.009	0.010
Inflation rate	−0.005***	−0.005***	−0.005***	−0.005***	−0.005***
Change in the natural logarithm of general government expenditure	0.035*	0.045**	0.037*	0.036*	0.035*
Change in the natural logarithm of government debt	−0.009	−0.003	−0.009	−0.005	−0.008
Change in the natural logarithm of the corruption perception index	0.021	0.029	0.025	0.026	0.023
Change in natural logarithm of trade (imports, exports)	0.027	0.020	0.031	0.024	0.027
The natural logarithm of economic inequality	−1.197	1.154	−0.959	−0.330	−0.791
The quadratic natural logarithm of economic inequality	0.182	−0.149	0.141	0.025	0.065
N (number of observations)	112	112	112	112	112
Adjusted R^2	0.765	0.750	0.758	0.760	0.768
F criteria (*p*-value)	<0.001	<0.001	<0.001	<0.001	<0.001
Breaking point	26.76	48.63	29.74	7.14	4.32
Actual range of the economic inequality indicator	from 20.9 to 40.8	from 37.2 to 61.6	from 24.3 to 46.8	from 4.49 to 18.07	from 3.03 to 8.32

Source: Authors' calculations

Note: Gini 1 – Gini coefficient of equivalized disposable income; Gini 2 – Gini coefficient of equivalized disposable income before social transfers (pensions included in social transfers) Gini coefficient of equivalized disposable income before social transfers (pensions included in social transfers); Gini 3 – Gini coefficient of equivalized disposable income before social transfers (pensions excluded from social transfers); DK – X/I deciles' differentiation coefficient; S80/S20 – income quintile ratio

* 90% of significance level.

** 95% of significance level.

*** 99% of significance level.

48.63% the marginal effect of economic inequality on economic growth becomes negative, i.e., economic inequality becomes excessive. In this way, according to official statistics for 2018, economic inequality is excessive and had a negative impact on the economic growth in more than half of the EU countries – Sweden (57.1%), Greece (57%), Portugal (56.5%), Germany (56.4%), Bulgaria (54.8%), Romania (54.6%), the United Kingdom (53.7%), Luxembourg (51.9%), Lithuania (51.1%), France (50.9%), Croatia (49.8%), Denmark (49%), Hungary (49.1%), Finland (48.8%), and Spain (48.7%).

Meanwhile, in the general context of the EU-28, the Gini coefficient of equivalized disposable income (Gini 1), the Gini coefficient of equivalized disposable income before social transfers, where pensions are eliminated from social transfers (Gini 3), the X/I deciles' coefficient of differentiation, and the quantile income S80/S20 have a negative marginal effect on four-year economic growth. The breaking point when the marginal effect of economic inequality on economic growth becomes positive is 26.76%, 29.74%, 7.14 and 4.32, respectively (see Table 3.19).

The breakdown of the EU-28 countries into countries with higher standard of life and to countries with lower standard of life and the Chow test reveals that excessive inequality in the less wealthy group of countries is observed when Gini coefficient of equivalized disposable income before social transfers, when pensions are included in social transfers (Gini 2) reaches breaking point at 46.15% (see Table 3.20). Therefore, according to the 2018 data, the economic inequality has had a negative impact on economic growth in Greece (57%), Portugal (56.5%), Bulgaria (54.8%), Romania (54.6%), Lithuania (51.1%), Croatia (49.8%), Hungary (49.1%), Latvia (48.1%), and Poland (46.3%).

In the group of richer countries, excessive inequality is evident by assessing economic inequality by Gini coefficient of equivalized disposable income before social transfers, when pensions are included in social transfers (Gini 2), and Gini coefficient of equivalized disposable income before social transfers, when pensions are eliminated from social transfers (Gini 3). The results show that the increase in economic inequality has a positive marginal effect on economic growth, i.e., greater economic inequality is linked to faster economic growth. But at a breaking point at 53.42% and 34.46%, respectively, the marginal effect of economic inequality on economic growth becomes negative and further increases in economic inequality are linked to slowing economic growth (see Table 3.20). As a result, according to statistics, in 2018, excessive inequality was recorded in Sweden (Gini 2 is 57.1%, Gini 3 is 35.7%), Germany (Gini 2 – 56.4%, Gini 3 – 36.6%), the United Kingdom (Gini 2 – 53.7%), Ireland (Gini 3 – 39.3%), Luxembourg (Gini 3 – 38.1%), Spain (Gini 3 – 37%), Denmark (Gini 3 –36%), Italy (Gini 3 – 35.7%), and France (Gini 3 – 34.9%).

The impact of excessive inequality on the average five-year growth rate. The analysis of the impact of economic inequality on the five-year rate of economic growth did not reveal a statistically significant indicator of

Table 3.20 The impact of economic inequality on socio-economic progress by group of countries, reflecting through economic growth indicators and controlling other growth factors (the dependent variable is average four-year economic growth rate)

	Coefficient estimates calculated using FEM				
	Indicators of economic inequality				
Variables	*Gini 1*	*Gini 2*	*Gini 3*	*DK*	*S80/S20*
Dependent variable	Average four-year economic growth rate				
	Less wealthy countries				
Constant	0.663	−2.220	0.046	0.472	0.249
Change in natural logarithm of GDP per capita ($t − 4$)	−0.071★★★	−0.068★★★	−0.066★★	−0.069★★	−0.070★★
Change in the natural logarithm of investments	−0.020	−0.021	−0.016	−0.022	−0.021
Change in the natural logarithm of the share of population with tertiary education level	0.022	0.021	0.021	0.017	0.019
Inflation rate	−0.003★★★	−0.003★★★	−0.003★★★	−0.003★★★	−0.003★★★
Change in the natural logarithm of general government expenditure	0.016	0.020	0.017	0.015	0.016
Change in the natural logarithm of government debt	0.001	0.007	0.003	0.005	0.003
Change in the natural logarithm of the corruption perception index	0.023	0.029	0.025	0.025	0.025
Change in natural logarithm of trade (imports, exports)	0.044★★	0.035★★	0.039★★	0.040★★	0.044★★
The natural logarithm of economic inequality	−0.184	1.362	0.134	−0.026	0.025
The quadratic natural logarithm of economic inequality	0.034	−0.178	−0.012	0.003	0.000
Breaking point	−	46.15	−	−	−
	The difference between rich and less rich countries				
The pseudo-variable of richer countries	7.948	−1.081	0.740	2.813	3.582
Change in natural logarithm of GDP per capita ($t − 4$)	−0.221★★★	−0.186★★★	−0.197★★★	−0.191★★★	−0.193★★★
Change in the natural logarithm of investments	0.017	0.007	0.005	0.011	0.013
Change in the natural logarithm of the share of population with tertiary education level	−0.003	−0.010	−0.008	−0.005	−0.007

(Continued)

Table 3.20 The impact of economic inequality on socio-economic progress by group of countries, reflecting through economic growth indicators and controlling other growth factors (the dependent variable is average four-year economic growth rate)   *(Continued)*

Variables	Coefficient estimates calculated using FEM				
	Indicators of economic inequality				
	Gini 1	Gini 2	Gini 3	DK	S80/S20
Inflation rate	−0.001	0.000	−0.001	−0.001	−0.001
Change in the natural logarithm of general government expenditure	−0.022	0.006	0.022	0.013	0.011
Change in the natural logarithm of government debt	−0.090***	−0.095***	−0.088***	−0.096***	−0.093***
Change in the natural logarithm of the corruption perception index	−0.065	−0.052	−0.038	−0.045	−0.045
Change in natural logarithm of trade (imports, exports)	−0.046***	−0.039**	−0.053***		−0.056	−0.058***
The natural logarithm of economic inequality	−2.819	1.834	1.155	−0.009	−0.235
The quadratic natural logarithm of economic inequality	0.412	−0.224	−0.170	−0.001	0.016
Breaking point	28.97	53.42	34.46	–	7.62
Actual range of the economic inequality indicator	from 20.9 to 40.8	from 37.2 to 61.6	from 24.3 to 46.8	from 4.49 to 18.07	from 3.03 to 8.32
N (number of observations)	112	112	112	112	112
Adjusted R^2	0.912	0.910	0.908	0.906	0.908
F criteria (p-value)	<0.001	<0.001	<0.001	<0.001	<0.001

Source: Authors' calculations.

Note: Gini 1 – Gini coefficient of equivalized disposable income; Gini 2 – Gini coefficient of equivalized disposable income before social transfers (pensions included in social transfers) Gini coefficient of equivalized disposable income before social transfers (pensions included in social transfers); Gini 3 – Gini coefficient of equivalized disposable income before social transfers (pensions excluded from social transfers); DK – X/I deciles' differentiation coefficient; S80/S20 – income quintile ratio.

* 90% of significance level.

** 95% of significance level.

*** 99% of significance level.

economic inequality, but the low significance of complex significance criterion (p-value) highlighted the high reliability of calculations and results obtained, which suggests that the identified factors have an impact on economic growth (see Table 3.21). According to estimates of excessive inequality, i.e., situation where the marginal effect of economic inequality on economic growth is negative, the assessment of economic inequality under Gini coefficient of equivalized disposable income before social transfers,

Table 3.21 The impact of economic inequality on socio-economic progress, reflecting economic growth indicators and controlling other growth factors (the dependent variable is average five-year economic growth rate)

	Coefficient estimates calculated using FEM				
	Indicators of economic inequality				
Variables	Gini 1	Gini 2	Gini 3	DK	S80/S20
Dependent variable	Average five-year economic growth rate				
Constant	0.929	−0.809	−0.457	1.260	1.678
Change in natural logarithm of GDP per capita ($t-5$)	−0.066***	−0.070***	−0.062**	−0.067**	−0.065**
Change in the natural logarithm of investments	−0.016	−0.017	−0.019	−0.014	−0.015
Change in the natural logarithm of the share of population with tertiary education level	0.001	−0.001	0.000	−0.003	−0.002
Inflation rate	−0.003***	−0.003***	−0.003***	−0.003***	−0.003***
Change in the natural logarithm of general government expenditure	0.013	0.018	0.014	0.014	0.015
Change in the natural logarithm of government debt	0.007	0.007	0.007	0.006	0.007
Change in the natural logarithm of the corruption perception index	0.026	0.031*	0.027	0.029*	0.025
Change in natural logarithm of trade (imports, exports)	0.046***	0.046***	0.047***	0.050***	0.047***
The natural logarithm of economic inequality	−0.344	0.627	0.426	−0.273	−0.423
The quadratic natural logarithm of economic inequality	0.050	−0.084	−0.060	0.020	0.033
N (number of observations)	84	84	84	84	84
Adjusted R^2	0.825	0.836	0.825	0.834	0.831
F criteria (*p*-value)	<0.001	<0.001	<0.001	<0.001	<0.001
Breaking point	30.78	40.95	34.72	9.74	6.00
Actual range of the economic inequality indicator	from 20.9 to 40.8	from 37.2 to 61.6	from 24.3 to 46.8	from 4.49 to 18.07	from 3.03 to 8.32

Source: Authors' calculations.

Note: Gini 1 – Gini coefficient of equivalized disposable income; Gini 2 – Gini coefficient of equivalized disposable income before social transfers (pensions included in social transfers) Gini coefficient of equivalized disposable income before social transfers (pensions included in social transfers); Gini 3 – Gini coefficient of equivalized disposable income before social transfers (pensions excluded from social transfers); DK – X/I deciles' differentiation coefficient; S80/S20 – income quintile ratio.

* 90% of significance level.

** 95% of significance level.

*** 99% of significance level.

when pensions are included in social transfers (Gini 2), and Gini coefficient of equivalized disposable income before social transfers, where pensions are eliminated from social transfers (Gini 3), reveals the breaking points of 40.95% and 34.72%, respectively. According to 2018 statistics, excessive inequality expressed by Gini 2, which has a negative impact on the economic growth, was recorded in all EU countries (except Slovakia, where Gini 2 was 37.2%), and under Gini 3 in more than half of the EU countries: Bulgaria (43.3%), Lithuania (40.6%), the United Kingdom (40.4%), Ireland (39.3%), Latvia (38.2%), Luxembourg (38.1%), Romania (38%), Spain (37%), Germany (36.6%), Denmark (36%), Sweden (35.7%), Italy (35.7%), Portugal (35.2%), Greece (35.2%), and France (34.9%).

The statistically significant X/I deciles' differentiation coefficient was obtained in the less rich countries when the EU-28 countries are divided into countries having the higher standard of life and countries having lower standard of life and the Chow test applied (see Table 3.22). In this way, it is clear that when assessing economic inequality in terms of the X/I deciles' differentiation coefficient, the marginal effect of economic inequality on five-year economic growth is negative for countries with the lower standard of life under 8.63 levels. By reaching this breaking point, the marginal effect on economic growth becomes a positive, as the growth of the income of the richest population group stimulates them to invest additionally in the improvement of the country's economy.

The results of the research lead to an important conclusion that economic growth depends not only on economic indicators (GDP volume, investment [gross fixed capital formation], share of population with tertiary education level, inflation rate, government expenditure, government debt level, corruption perception index, volume of trade [import, export]), but also on economic inequality indicators such as Gini coefficient of equivalized disposable income (Gini 1), Gini coefficient of equivalized disposable income before social transfers, when pensions are included in social transfers (Gini 1), Gini coefficient of equivalized disposable income before social transfers, pensions eliminated from social transfers (Gini 3), X/I deciles' differentiation coefficient, and S80/S20 income quintile ratio.

The low significance of the criterion F (*p*-value) of the models demonstrates, the high reliability of the calculations and results obtained and suggests:

- firstly, that economic inequality affects the economic growth of the EU-28 countries;
- secondly, the impact of economic inequality on economic growth is evident over the longer term;
- thirdly, although the statistical significance of economic inequality indicators is already occurring in the current year of economic growth, it is particularly evident in analysis of economic inequality impact on the average economic growth rate of the two- and three-year period; and

Table 3.22 The impact of economic inequality on socio-economic progress by group of countries, reflecting through economic growth indicators and controlling other economic growth factors (the dependent variable is average five-year economic growth rate)

Variables	Coefficient estimates calculated using FEM				
	Indicators of economic inequality				
	Gini 1	Gini 2	Gini 3	DK	S80/S20
Dependent variable	Average four-year economic growth rate				
	Less wealthy countries				
Constant	2.131	−1.854	1.429	1.625	1.169
Change in natural logarithm of GDP per capita ($t-5$)	−0.064**	−0.075***	−0.060**	−0.090***	−0.067**
Change in the natural logarithm of investments	−0.004	−0.005	−0.006	0.002	−0.003
Change in the natural logarithm of the share of population with tertiary education level	0.006	−0.001	0.001	0.000	−0.003
Inflation rate	−0.002**	−0.002	−0.002**	−0.002**	−0.002**
Change in the natural logarithm of general government expenditure	0.010	0.018	0.011	0.009	0.013
Change in the natural logarithm of government debt	0.008	0.008	0.008	0.006	0.007
Change in the natural logarithm of the corruption perception index	0.022	0.025*	0.023	0.028*	0.021
Change in natural logarithm of trade (imports, exports)	0.052**	0.048**	0.055**	0.042*	0.049*
The natural logarithm of economic inequality	−0.240	0.234	0.320	−0.400*	−0.340
The quadratic natural logarithm of economic inequality	0.035	−0.038	−0.046	0.030*	0.026
Breaking point	30.73	−	33.41	8.63	6.31
	The difference between rich and less rich countries				
The pseudo-variable of richer countries	−	−	−	−	−
Change in natural logarithm of GDP per capita ($t-5$)	−0.069	−0.059	−0.071	−0.035	−0.047
Change in the natural logarithm of investments	−0.090*	−0.088**	−0.084*	−0.091**	−0.089*

(Continued)

Table 3.22 The impact of economic inequality on socio-economic progress by group of countries, reflecting through economic growth indicators and controlling other economic growth factors (the dependent variable is average five-year economic growth rate)   *(Continued)*

| Variables | Coefficient estimates calculated using FEM | | | | |
| | Indicators of economic inequality | | | | |
	Gini 1	Gini 2	Gini 3	DK	S80/S20
Change in the natural logarithm of the share of population with tertiary education level	0.016	0.016	0.019	0.010	0.016
Inflation rate	0.003	0.004**	0.003	0.003	0.003*
Change in the natural logarithm of general government expenditure	−0.146*	−0.134**	−0.133*	−0.075	−0.103
Change in the natural logarithm of government debt	−0.055**	−0.059***	−0.057**	−0.062***	−0.059**
Change in the natural logarithm of the corruption perception index	−0.091	−0.098	−0.083	−0.047	−0.060
Change in natural logarithm of trade (imports, exports)	−0.073*	−0.060	−0.076*	−0.049	−0.062
The natural logarithm of economic inequality	−0.528	3.371	−0.936	0.683	0.815
The quadratic natural logarithm of economic inequality	0.076	0.421	0.128	−0.052	−0.067
Breaking point	31.47	50.85	41.41	6.21	3.62
Actual range of the economic inequality indicator	from 20.9 to 40.8	from 37.2 to 61.6	from 24.3 to 46.8	from 4.49 to 18.07	from 3.03 to 8.32
N (number of observations)	84	84	84	84	84
Adjusted R^2	0.902	0.923	0.902	0.911	0.904
F criteria (p-value)	<0.001	<0.001	<0.001	<0.001	<0.001

Source: Authors' calculations.

Note: Gini 1 – Gini coefficient of equivalized disposable income; Gini 2 – Gini coefficient of equivalized disposable income before social transfers (pensions included in social transfers)Gini coefficient of equivalized disposable income before social transfers (pensions included in social transfers); Gini 3 – Gini coefficient of equivalized disposable income before social transfers (pensions excluded from social transfers); DK – X/I deciles' differentiation coefficient; S80/S20 – income quintile ratio.

* 90% of significance level.

** 95% of significance level.

*** 99% of significance level.

- fourthly, the impact of economic inequality on economic growth varies from country to country, depending on living standards of these countries.

In the current year, in richer countries, the increase in economic inequality has a positive marginal effect on economic growth, but with economic inequality reaching 34.03% of Gini coefficient of equivalized disposable income before social transfers, when pensions are eliminated from social transfers (Gini 3), economic inequality turns into excessive inequality when the marginal effect of economic inequality on economic growth becomes negative. Further increase in economic inequality expressed by Gini 3 coefficient is linked to slowing economic growth.

When analyzing the impact of economic inequality on the average economic growth rates of the two-year period in the group of richer countries, the increase in economic inequality has a positive marginal effect on economic growth and, at a breaking point of 34.98% in terms Gini coefficient of equivalized disposable income before social transfers, pensions are eliminated from social transfers (Gini 3), economic inequality becomes excessive inequality and the marginal effect of it on economic growth becomes negative, i.e., further increase in economic inequality is linked to slowing economic growth.

When analyzing the impact of economic inequality on the average three-year economic growth rate, the marginal effect of economic inequality on economic growth in the wealthier group of countries is positive and, just with Gini 1 coefficient reaching 29.63%, Gini 2 coefficient − 53.06%, and Gini 3 coefficient − 36.45%, economic inequality becomes excessive and the marginal effect of economic inequality on economic growth becomes negative, i.e., further growth of economic inequality is linked to slowing economic growth.

By analyzing the impact of economic inequality on the average economic growth rates of two- and three-year periods, the trajectory of the inverted "U" curve can be observed in a group of less-wealthy countries, where the marginal effect of economic inequality on economic growth is negative and only when the breaking point is reached, the marginal effect on economic growth becomes positive. In this way, it can be assumed that the richest members of society contribute to the country's well-being as their income increases, which influences the appropriate direction of economic growth of less wealthy countries.

Despite the fact that the assessment of the impact of economic inequality on the average economic growth rates of the four- and five-year economics did not reveal a statistically significant indicator of economic inequality, the low significance of the criterion F (p-value) of the complex significance of indicators shows the high reliability of the calculations and results obtained. So, all the identified factors, including economic inequality indicators, in the complex have an impact on economic growth in the long term.

According to the research, excessive inequality, which has a negative impact on economic growth, are especially evident in richer countries. Economies in less wealthy countries require income growth, even if it exacerbates economic inequality and widens the gap between the richest and poorest members of society. The research shows that economic inequality is excessive and have a negative impact on the economies of richest countries such as Ireland, Denmark, Estonia, Italy, Spain, the United Kingdom, Luxembourg, France, Finland, Sweden, and Germany.

3.5 Negative consequences of excessive inequality on quality of life

The results of the empirical research on the impact of economic inequality on quality of life are presented in the following sequence:

1 the impact of economic inequality on quality of life by the LSM and the method of standard errors correcting heteroskedasticity;
2 the impact of economic inequality on quality of life according to the healthy life years; and
3 the impact of economic inequality on quality of life, the latter being expressed as an indicator of emigration rate.

Assessment of economic inequality impact on quality of life by the LSM. Measuring quality of life by the median income indicator and assessing the impact of economic inequality on quality of life by the LSM, the hypothesis (H0) is tested that all the indicators highlighted do not affect quality of life. Based on the results of which are presented in Table 3.23, the complex criterion of the significance of indicators F (p-value) is low, demonstrating the high reliability of the calculations and results obtained, which results in a hypothesis rejected.

Table 3.23 The impact of economic inequality on socio-economic progress, reflecting through the quality-of-life indicators (the dependent variable is logarithmic median income)

Variables	Coefficient estimates calculated using SSA				
	Indicators of economic inequality				
	Gini 1	Gini 2	Gini 3	DK	S80/S20
Dependent variable	Quality of life (logarithmic median income)				
Constant	−61.509***	−172.19***	−57.110***	−68.145***	−86.717***
Change in natural logarithm of life expectancy	11.373***	12.449***	12.192***	11.396***	10.955***
Change in the natural logarithm of the poverty level	−0.389***	−0.444***	−0.769***	−0.355***	−0.341**
The change in the natural logarithm of a healthy life year	−1.120***	−1.049***	−1.100***	−1.031***	−1.076***

(Continued)

Table 3.23 The impact of economic inequality on socio-economic progress, reflecting through the quality-of-life indicators (the dependent variable is logarithmic median income)   *(Continued)*

	Coefficient estimates calculated using SSA				
	Indicators of economic inequality				
Variables	*Gini 1*	*Gini 2*	*Gini 3*	*DK*	*S80/S20*
Change in the natural logarithm of the average expected years of schooling	0.552**	0.300	0.278	0.616	0.574
Change in the natural logarithm of the unemployment rate	−0.091*	−0.086*	−0.130***	−0.105	−0.087
Change in the natural logarithm of the emigration rate	−0.007	−0.042**	−0.034**	−0.011	−0.004
Change in the natural logarithm of the severe housing deprivation rate	−0.096	−0.100***		−0.064	−0.100***	−0.100***
Change in the natural logarithm of government spending on public order and safety	−0.707	−0.513***	−0.462	−0.700***	−0.717***
The natural logarithm of economic inequality	15.071*	67.290***	9.605	9.413***	16.934***
The quadratic natural logarithm of economic inequality	−2.168*	−8.510***	−1.109	−0.689***	−1.357***
N (number of observations)	218	218	218	218	218
Adjusted R^2	0.888	0.897	0.905	0.894	0.892
F criteria (p-value)	<0.001	<0.001	<0.001	<0.001	<0.001
The value of test p to check for autocorrelation	0.000	0.000	0.000	0.000	0.000
Value of test p to check heteroskedasticity	0.009	0.016	0.015	0.017	0.007
Breaking point	32.30	52.13	–	9.30	5.13
Actual range of the economic inequality indicator	from 20.9 to 40.8	from 37.2 to 61.6	from 24.3 to 46.8	from 4.49 to 18.07	from 3.03 to 8.32

Source: Authors' calculations.

Note: Gini 1 – Gini coefficient of equivalized disposable income; Gini 2 – Gini coefficient of equivalized disposable income before social transfers (pensions included in social transfers)Gini coefficient of equivalized disposable income before social transfers (pensions included in social transfers); Gini 3 – Gini coefficient of equivalized disposable income before social transfers (pensions excluded from social transfers); DK – X/I deciles' differentiation coefficient; S80/S20 – income quintile ratio.

* 90% of significance level.

** 95% of significance level.

*** 99% of significance level.

Therefore, the opposite is true; the indicators distinguished (together with indicators of economic inequality) have an impact on quality of life.

Based on the calculations made (see Table 3.23), statistically significant indicators of economic inequality are: Gini coefficient of equivalized disposable income (Gini 1) and Gini coefficient of equivalized disposable income before social transfers, pensions included in social transfers (Gini 2). Economic inequality expressed by Gini 1 reveals that up to the identified breaking point (32.30%), the increase in economic inequality has a positive marginal effect on quality of life, and then turns into excessive inequality, when the marginal effect of economic inequality on quality of life becomes negative and the further increase in economic inequality is linked to a deterioration in quality of life (see Figure 3.20). According to 2018 data, excessive inequality was recorded in Bulgaria (Gini 1 was 39.6%), Lithuania (36.9%), Latvia (35.6%), Romania (35.1%), the United Kingdom (33.5%), Italy (33.4%), Spain (33.2%), Luxembourg (33.2%), and Greece (32.3%).

When economic inequality is measured by Gini 2, the marginal effect is positive and increasing economic inequality has positive impact on the

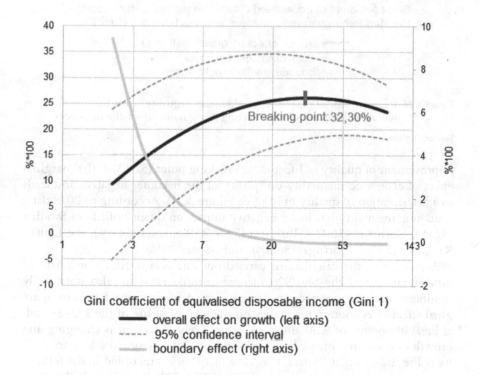

Figure 3.20 Relative slope of the impact of economic inequality (according to Gini 1) on socio-economic progress, reflecting through quality-of-life indicators.

Source: Authors' calculations.

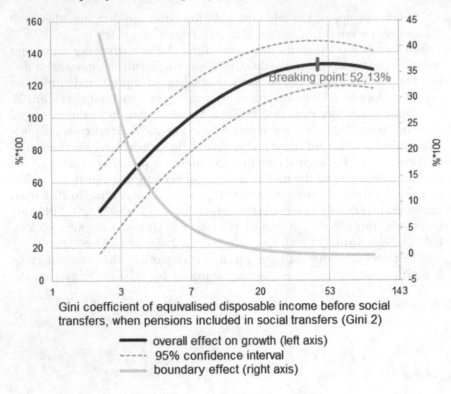

Figure 3.21 Relative slope of the impact of economic inequality (according to Gini 2) on socio-economic progress, reflecting through quality-of-life indicators.

Source: Authors' calculations.

improvement of quality of life and, at breaking point (52.13%), the marginal effect of economic inequality on quality of life becomes negative and leads to a deterioration of quality of life (see Figure 3.21). According to 2018 data, economic inequality has had a negative impact on quality of life in Sweden (57.1%), Germany (56.4%), Bulgaria (54.8%), Romania (54.6%), the United Kingdom (53.7%), Portugal (56.5%), and Greece (57%).

According to the calculations carried out, the X/I deciles' coefficient of differentiation and the S80/S20 income quintile ratio are also statistically significant indicators of economic inequality, demonstrating a positive marginal effect of economic inequality on the quality of life of the EU-28 and, at breaking points of 9.30 and 5.13, respectively, direction is changing and growth of economic inequality results in a negative marginal effect on quality of life. Based on 2018 data, excessive inequality is recorded in the following countries: Bulgaria (X/I – 14.9, S80/S20 – 7.7), Romania (X/I – 14.4, S80/S20 – 7.2), Lithuania (X/I – 13.1, S80/S20 – 7.1), Italy (X/I – 12.6, S80/S20 – 6.1), Latvia (X/I – 12.4, S80/S20 – 6.8), Spain (X/I – 11.0, S80/S20 – 6.0), Luxembourg (X/I – 10.3, S80/S20 – 6.0), the United Kingdom

(X/I – 10.0, S80/S20 – 5.6), Germany (X/I – 9.9), Greece (X/I – 9.8, S80/ S20 – 5.5), and Portugal (S80/S20 – 5.2).

The Chow test examines the hypothesis (H0), that the impact of economic inequality on quality of life in different groups of countries – countries with higher living standards (where GDP per capita in PPS is more than €24750) and in countries with lower living standards (where GDP per capita under PPS is less than €24750) – is the same in both groups. According to the results obtained, based on the Chow test, the complex significance criterion of the indicators F (*p*-value) is very low, which proves the high reliability of the calculations, which results in the hypothesis being rejected. Therefore, the impact of all excluded factors, including economic inequality, on quality of life varies between the groups of countries.

In a less wealthy group of countries, economic inequality is statistically significant, with only one of the indicators – the Gini coefficient of equivalized disposable income before social transfers, pensions included in social transfers (Gini 2). The results show that in a less wealthy group of countries, the increase in economic inequality has a positive marginal effect on quality of life and, when Gini 2 reaches 43.25%, the marginal effect of economic inequality on quality of life becomes negative and the further increase in economic inequality is linked to the deteriorating quality of life. In this way, in 2018 economic inequality negatively affected quality of life in Greece (57%), Portugal (56.5%), Bulgaria (54.8%), Romania (54.6%), Lithuania (51.1%), Croatia (49.8%), Hungary (49.1%), Latvia (48.1%), and Poland (46.3%).

In richer countries, all indicators of economic inequality are statistically significant – the Gini coefficient of equivalized disposable income (Gini 1); the Gini coefficient of equivalized disposable income before social transfers, in the cases of pensions included in social transfers (Gini 2); the Gini coefficient of equivalized disposable income before social transfers, where pensions are eliminated from social transfers (Gini 3); X/I deciles' coefficient of differentiation; and S80/S20 income quintile ratio. In richer countries, the increase in economic inequality has a positive marginal effect on quality of life and economic inequality becomes excessive when Gini 1 reaches the level of 29.16%, Gini 2 – 52.95%, Gini 3 – 44.48, X/I – 8.46, and S80/S20 – 4.94. After this breaking point, economic inequality have a negative impact on quality of life. According to 2018 data, the negative impact of economic inequality on quality of life was observed in the United Kingdom (Gini 1 – 33.5%, Gini 2 – 53.7%, X/I – 10.0, S80/S20 – 5.6), Italy (Gini 1 – 33.4%, X/ I – 12.6, S80/S20 – 6.1), Spain (Gini 1 – 33.2%, X/I – 11.0, S80/S20 – 6.0), Luxembourg (Gini 1 – 33.2%, X/I – 10.3, S80/S20 – 5.7), Germany (Gini 1 – 31 Gini 2 – 56.4%, X/I – 9.9, S80/S20 – 5.1), Estonia (Gini 1 – 30.6%), and Sweden (Gini 2 – 57.1%).

The reliability of the LSM model was tested. First, Wooldridge's test for autocorrelation determination was applied, when the hypothesis (H0) is tested, that for the model the autocorrelation is characteristic and the obtained test *p*-values <0.001 (see Table 3.24) indicates that there is autocorrelation

Table 3.24 The impact of economic inequality on socio-economic progress by group of countries, reflecting through the quality-of-life indicators (dependent variable is logarithmic median income)

| Variables | Coefficient estimates calculated using SSA | | | | |
| | Indicators of economic inequality | | | | |
	Gini 1	Gini 2	Gini 3	DK	S80/S20
Dependent variable	Quality of life (logarithmic median income)				
	Less wealthy countries				
Constant	−27.289**	−97.815***	−47.103***	−29.822***	−41.320***
Change in natural logarithm of life expectancy	8.373***	9.418***	7.955***	8.168***	8.028***
Change in the natural logarithm of the poverty level	−0.586***	−0.184	0.669***	−0.448***	−0.495***
The change in the natural logarithm of a healthy life year	−0.298	−0.425	−0.214	−0.276	−0.247
Change in the natural logarithm of the average expected years of schooling	0.202	0.608	0.106	0.368	0.335
Change in the natural logarithm of the unemployment rate	0.190***	0.285***	0.214***	0.196***	0.195***
Change in the natural logarithm of the emigration rate	−0.067***	−0.053***	−0.069***	−0.065***	−0.066***
Change in the natural logarithm of the severe housing deprivation rate	−0.010	0.006	−0.017	−0.026	−0.021
Change in the natural logarithm of government spending on public order and safety	0.552***	−0.764***	−0.533***	−0.575***	−0.587***
The natural logarithm of economic inequality	1.452	35.632***	13.536	1.621	5.659
The quadratic natural logarithm of economic inequality	−0.165	−4.730***	−1.839	−0.120	−0.451
Breaking point	–	43.25	39.64	8.70	5.30
The difference between rich and less rich countries					
The pseudo-variable of richer countries	−115.22***	−237.17***	−85.244***	−74.790***	−145.26***
Change in natural logarithm of life expectancy	1.633	3.474	5.238	3.790	3.325
Change in the natural logarithm of the poverty level	0.472*	−0.063	−0.093	0.246	0.069

(Continued)

Table 3.24 The impact of economic inequality on socio-economic progress by group of countries, reflecting through the quality-of-life indicators (dependent variable is logarithmic median income) *(Continued)*

	Coefficient estimates calculated using SSA				
	Indicators of economic inequality				
Variables	Gini 1	Gini 2	Gini 3	DK	S80/S20
The change in the natural logarithm of a healthy life year	−0.879*	−0.495	−0.946**	−0.945**	−1.001**
Change in the natural logarithm of the average expected years of schooling	0.333	−0.209	0.382	0.326	0.531
Change in the natural logarithm of the unemployment rate	−0.246**	−0.607***	−0.512***	−0.267***	−0.213**
Change in the natural logarithm of the emigration rate	0.100***	−0.024	0.016	0.068**	0.065*
Change in the natural logarithm of the severe housing deprivation rate	−0.048	−0.077**	−0.001	−0.041	−0.040
Change in the natural logarithm of government spending on public order and safety	−0.116	0.506**	0.336*	−0.039	0.005
The natural logarithm of economic inequality	65.017***	112.073***	34.614**	17.950***	42.983***
The quadratic natural logarithm of economic inequality	−9.689***	−13.876***	−4.505**	−1.332***	−3.470***
Breaking point	29.16	52.95	44.48	8.46	4.94
Actual range of the economic inequality indicator	from 20.9 to 40.8	from 37.2 to 61.6	from 24.3 to 46.8	from 4.49 to 18.07	from 3.03 to 8.32
N (number of observations)	212	212	212	212	212
Adjusted R^2	0.936	0.953	0.949	0.939	0.940
F criteria (p-value)	<0.001	<0.001	<0.001	<0.001	<0.001

Source: Authors' calculations.

Note: Gini 1 − Gini coefficient of equivalized disposable income; Gini 2 − Gini coefficient of equivalized disposable income before social transfers (pensions included in social transfers), Gini coefficient of equivalized disposable income before social transfers (pensions included in social transfers); Gini 3 − Gini coefficient of equivalized disposable income before social transfers (pensions excluded from social transfers); DK − X/I deciles' differentiation coefficient; S80/S20 − income quintile ratio.

* 90% of significance level.

** 95% of significance level.

*** 99% of significance level.

in this model. Secondly, the White test to determine heteroskedicity was applied by testing the hypothesis (H0) that the model was characterized by heteroskedicity and the mean p-value 0.013 obtained (see Table 3.24), confirms the existence of heteroskedicity. Therefore, in order to provide a more reliable model for assessing the impact of economic inequality on quality of life, as well as addressing the problems of autocorrelation and heteroskedism, the model will be recalculated using the standard error method that corrects heteroskedicity.

The assessment of impact of economic inequality on quality of life by applying the heteroskedicity corrected standard error method (HCSEM). Measuring the quality of life by the median income indicator and assessing the impact of economic inequality on quality of life by the HCSEM on the hypothesis (H0) that all excluded factors do not affect quality of life are tested. Based on the results of which are presented in Table 3.25, the complex significance criterion of the indicators F (p-value) is low, demonstrating the high reliability of the calculations and results obtained, which results in a rejection of the hypothesis and states that the indicators identified in the model have an impact on quality of life.

When assessing the impact of economic inequality on quality of life according to the HCSEM, all indicators of economic inequality are statistically significant, but the most pronounced excessive inequality is evident in the assessment of economic inequality by the Gini coefficient of equivalized disposable income (Gini 1), the X/I deciles' coefficient of differentiation, and the 80/S20 income quintile ratio. The results show that the increase in economic inequality has a positive marginal effect on the quality of life of EU countries until the following breaking points of economic inequality indicators are reached: 29.24%, 7.92, and 4.64, respectively. Then, the marginal

Table 3.25 The impact of economic inequality on socio-economic progress, reflecting through quality-of-life indicators (dependent variable is logarithmic median income)

	Coefficient estimates calculated using HCSEM				
	Indicators of economic inequality				
Variables	Gini 1	Gini 2	Gini 3	DK	S80/S20
Dependent variable	Quality of life (logarithmic median income)				
Constant	−55.453***	−183.02***	−59.838***	−59.731***	−76.020***
Change in natural logarithm of life expectancy	10.012***	11.744***	12.455***	9.800***	9.449***
Change in the natural logarithm of the poverty level	−0.308***	−0.262***	−0.725***	−0.308***	−0.301***

(Continued)

Table 3.25 The impact of economic inequality on socio-economic progress, reflecting through quality-of-life indicators (dependent variable is logarithmic median income)   *(Continued)*

	Coefficient estimates calculated using HCSEM				
	Indicators of economic inequality				
Variables	Gini 1	Gini 2	Gini 3	DK	S80/S20
The change in the natural logarithm of a healthy life year	−0.886***	−1.038***	−1.254***	−0.868***	−0.919
Change in the natural logarithm of the average expected years of schooling	0.020	−0.042	−0.002	0.196	0.072
Change in the natural logarithm of the unemployment rate	−0.023	−0.073*	−0.152***	−0.002	−0.003
Change in the natural logarithm of the emigration rate	0.010	−0.017	−0.029**	0.018	0.021*
Change in the natural logarithm of the severe housing deprivation rate	−0.112***	−0.114***	−0.065***	−0.114***	−0.119***
Change in the natural logarithm of government spending on public order and safety	−0.907***	−0.823***	−0.561***	−0.910***	−0.889***
The natural logarithm of economic inequality	15.461**	74.793***	11.006**	9.177***	15.944***
The quadratic natural logarithm of economic inequality	−2.290**	9.508***	−1.271*	−0.687***	−1.298***
N (number of observations)	218	218	218	218	218
Adjusted R^2	0.936	0.938	0.948	0.942	0.940
F criteria (p-value)	<0.001	<0.001	<0.001	<0.001	<0.001
Breaking point	29.24	–	–	7.92	4.64
Actual range of the economic inequality indicator	from 20.9 to 40.8	from 37.2 to 61.6	from 24.3 to 46.8	from 4.49 to 18.07	from 3.03 to 8.32

Source: Authors' calculations.

Note: Gini 1 – Gini coefficient of equivalized disposable income; Gini 2 – Gini coefficient of equivalized disposable income before social transfers (pensions included in social transfers); Gini coefficient of equivalized disposable income before social transfers (pensions included in social transfers); Gini 3 – Gini coefficient of equivalized disposable income before social transfers (pensions excluded from social transfers); DK – X/I deciles' differentiation coefficient; S80/S20 – income quintile ratio.

* 90% of significance level.

** 95% of significance level.

*** 99% of significance level.

effect of economic inequality on quality of life becomes negative, leading to a deterioration in quality of life in the EU–28 (see Table 3.25). Based on 2018 data, excessive inequality was recorded in almost half of the EU countries: Bulgaria (Gini 1 – 39.6%, X/I – 14.9, S80/S20 – 7.7), Lithuania (Gini 1 – 36.9%, X/I – 13.1, S80/S20 – 7.1), Latvia (Gini 1 – 35.6%, X/I – 12.4, S80/S20 – 6.8), Romania (Gini 1 – 35.1%, X/I – 14.4, S80/S20 – 7.2), the United Kingdom (Gini 1 – 33.5%, X/I – 10.0, S80/S20 – 5.6), Italy (Gini 1 – 33.4%, X/I – 12.6, S80/S20 – 6.1), Spain (Gini 1 – 33.2%, X/I – 11.0, S80/S20 – 6.0), Luxembourg (Gini 1 – 33.2%, X/I – 10.3, S80/S20 – 5.7), Greece (Gini 1 – 32.3%, X/I – 9.8, S80/S20 – 5.5), Portugal (Gini 1 – 32.1%, X/I – 8.7, S80/S20 – 5.2), Germany (Gini 1 – 31.1%, X/I – 9.9, S80/S20 – 5.1), Estonia (Gini 1 – 30.6, S80/S20 – 5.1), and Croatia (Gini 1 – 29.7%, X/I – 8.5, S80/S20 – 5.0).

By dividing the EU-28 countries into countries having higher standard of living and countries having lower living standards and applying the Chow test, the results showed that in richer countries, all indicators of economic inequality are statistically significant – the Gini coefficient of equivalized disposable income (Gini 1); the Gini coefficient of equivalized disposable income before social transfers, in the case of pensions included in social transfers (Gini 2); the Gini coefficient of equivalized disposable income before social transfers, where pensions are eliminated from social transfers (Gini 3); X/I deciles' coefficient of differentiation; and S80/S20 income quintile ratio (see Table 3.26). In richer countries, the increase in economic inequality has a positive marginal effect on quality of life and becomes excessive (the marginal effect negative impacted on quality of life is reached) when Gini 1 reaches the level of 28.87%, Gini 2 – 52.17%, Gini 3 – 41.00, X/I – 8.18, and S80/S20 – 4.90. In this way in 2018, economic inequality had a negative impact on quality of life in the United Kingdom (Gini 1 – 33.5%, Gini 2 – 53.7%, X/I – 10.0, S80/S20 – 5.6), Italy (Gini 1 – 33.4%, X/I – 12,6, S80/S20 – 6.1), Spain (Gini 1 – 33.2%, X/I – 11.0, S80/S20 – 6.0), Luxembourg (Gini 1 – 33.2%, X/I – 10.3, S80/S20 – 5.7), Germany (Gini 1 – 31.1%, Gini 2 – 56.4%, X/I – 9.9, S80/S20 – 5.1), Estonia (Gini 1 – 30.6%, S80/S20 – 5.1), Ireland (Gini 1 – 28.9%), and Sweden (Gini 2 – 57.1%).

In a less wealthy group of countries, economic inequality is statistically significant, with only one of the indicators of economic inequality identified – the Gini coefficient of equivalized disposable income before social transfers, pensions included in social transfers (Gini 2). The results show that in a less wealthy group of countries, the increase in economic inequality has a positive marginal effect on quality of life and, when Gini 2 reaches 40.86%, the marginal effect of economic inequality on quality of life becomes negative and the further increase in economic inequality is linked to a deteriorating quality of life. In this way, in 2018 economic inequality negatively affected quality of life in Greece (57%), Portugal (56.5%), Bulgaria (54.8%), Romania (54.6%), Lithuania (51.1%), Croatia (49.8%), Hungary (49.1%), Latvia (48.1%), and Poland (46.3%).

Table 3.26 The impact of economic inequality on socio-economic progress by group of countries, reflecting through quality-of-life indicators (dependent variable logarithmic income median)

Variables	Gini 1	Gini 2	Gini 3	DK	S80/S20
	— *Coefficient estimates calculated using HCSEM*				
	— *Indicators of economic inequality*				
Dependent variable	Quality of life (logarithmic median income)				
	Less wealthy countries				
Constant	−26.759***	−62.703***	−36.808***	−27.619***	−33.691***
Change in natural logarithm of life expectancy	7.830***	9.210***	8.154***	8.640***	8.310***
Change in the natural logarithm of the poverty level	−0.681***	−0.133*	−0.668***	−0.380***	−0.426***
The change in the natural logarithm of a healthy life year	−0.166	−0.865***	−0.011	−0.689***	−0.593***
Change in the natural logarithm of the average expected years of schooling	0.875***	0.700***	0.831***	0.544**	0.745***
Change in the natural logarithm of the unemployment rate	0.087**	0.128***	0.066*	0.007	0.021
Change in the natural logarithm of the emigration rate	−0.067***	−0.027**	−0.073***	−0.050***	−0.056
Change in the natural logarithm of the severe housing deprivation rate	0.034**	−0.022	−0.007	−0.018	−0.012
Change in the natural logarithm of government spending on public order and safety	−0.457***	−0.620***	−0.309***	−0.457***	−0.396***
The natural logarithm of economic inequality	1.198	18.354***	5.568	0.587	2.772
The quadratic natural logarithm of economic inequality	−0.113	−2.473***	−0.703	−0.038	−0.215
Breaking point	−	40.86	52.58	−	6.28
	The difference between rich and less rich countries				
The pseudo-variable of richer countries	−50.024	−129.794**	−74.239***	−11.835	−55.397**
Change in natural logarithm of life expectancy	−7.898***	−8.465***	−3.392***	−7.247***	−7.620***

(Continued)

Table 3.26 The impact of economic inequality on socio-economic progress by group of countries, reflecting through quality-of-life indicators (dependent variable logarithmic income median)   *(Continued)*

	Coefficient estimates calculated using HCSEM				
	Indicators of economic inequality				
Variables	Gini 1	Gini 2	Gini 3	DK	S80/S20
Change in the natural logarithm of the poverty level	0.296*	−0.279***	0.186	0.012	−0.177
The change in the natural logarithm of a healthy life year	0.421	0.979***	−0.488	0.618**	0.572*
Change in the natural logarithm of the average expected years of schooling	−0.479	−0.528*	−0.833***	−0.089	−0.136
Change in the natural logarithm of the unemployment rate	−0.077	−0.237***	−0.233***	0.006	0.022
Change in the natural logarithm of the emigration rate	0.059**	−0.024	0.021	0.038***	0.034*
Change in the natural logarithm of the severe housing deprivation rate	−0.035*	−0.003	−0.014	0.004	0.009
Change in the natural logarithm of govern-ment spending on public order and safety	0.093	0.401***	0.039	0.090	0.070
The natural logarithm of economic inequality	50.137***	82.994***	50.834***	12.454***	28.419***
The quadratic natural logarithm of economic inequality	−7.520***	−10.341***	−6.891***	−0.934***	−2.303***
Breaking point	28.87	52.17	41.00	8.18	4.90
Actual range of the economic inequality indicator	from 20.9 to 40.8	from 37.2 to 61.6	from 24.3 to 46.8	from 4.49 to 18.07	from 3.03 to 8.32
N (number of observations)	218	218	218	218	218
Adjusted R^2	0.986	0.982	0.989	0.989	0.991
F criteria (p-value)	<0.001	<0.001	<0.001	<0.001	<0.001

Source: Authors' calculations.

Note: Gini 1 – Gini coefficient of equivalized disposable income; Gini 2 – Gini coefficient of equivalized disposable income before social transfers (pensions included in social transfers); Gini 3 – Gini coefficient of equivalized disposable income before social transfers (pensions excluded from social transfers); DK – X/I deciles' differentiation coefficient; S80/S20 – income quintile ratio.

* 90% of significance level.

** 95% of significance level.

*** 99% of significance level.

The impact of economic inequality on the healthy life years. The statistically significant indicators of the economic inequality are Gini 1 (Gini coefficient of equivalized disposable income), Gini 2 (Gini coefficient of equivalized disposable income before social transfers, when pensions are included social transfers), and Gini 3 (Gini coefficient of equivalized disposable income before social transfers, when pensions are eliminated from social transfers). Assessing economic inequality by Gini coefficient of equivalized disposable income (Gini 1), the breaking point (25.34%) was identified. The increase in economic inequality has a positive marginal effect on the quality of life of EU countries, i.e., greater economic inequality is linked to the improvement of quality of life; but when the breaking point is reached, the marginal effect of economic inequality on quality of life becomes negative and the further increase in economic inequality is linked to the deterioration of the quality of life of EU countries (see Table 3.27). According to the result obtained, economic inequality have had a negative impact on the quality of life of almost all EU countries in 2018, as Gini 1 coefficient was more than 25.34% in all EU countries excluding the Czech Republic (24%), Slovenia (23.4%), and Slovakia (20.9%).

When assessing economic inequality based on in the Gini coefficient of equivalized disposable income before social transfers, where pensions are eliminated from social transfers (Gini 3), the marginal effect has a negative impact on the quality of life of EU countries to a break point at the 34.45% threshold, and then changes direction, and the marginal effect of economic inequality on quality of life becomes positive (see Table 3.27). The different directions achieved by measuring income inequality according to Gini 1 and Gini 3 are due to state social security policies, i.e., the Gini 3 coefficient assesses the income of the population along with pensions, but without other social benefits. Therefore, without the latter, the negative impact of economic inequality on quality of life, measured by healthy life years of life is observed.

By dividing the EU-28 countries into countries having the higher standard of life and countries having the lower standards of life and following the Chow test, the results showed that statistically significant values in the economic inequality were only recorded in the richer group of countries. Although the low significance criterion F (*p*-value) of the Chow test shows that the indicators excluded affect the quality of life in the EU-28 and their difference in different groups of countries. In a richer group of countries, the Gini coefficient of equivalized disposable income before social transfers when pensions are included in social transfers (Gini 2) and the X/I deciles' coefficient of differentiation are statistically significant (see Table 3.28).

Measured economic inequality by Gini 2 coefficient, the results show that in a group of richer countries, the marginal effect of economic inequality on quality of life is negative, and then Gini 2 coefficient reaches 50.19%, the marginal effect of economic inequality on quality of life becomes positive (see Table 3.28). Bearing in mind that the indicator of economic inequality

168   *The Impact of Excessive Inequality*

Table 3.27 Impact of economic inequality on socio-economic progress, reflecting quality-of-life indicators (dependent variable is logarithmic indicator for healthy life years)

| Variables | Coefficient estimates calculated using HCSEM | | | | |
| | Indicators of economic inequality | | | | |
	Gini 1	Gini 2	Gini 3	DK	S80/S20
Dependent variable	Quality of life (logarithmic indicator of healthy life years)				
Constant	−6.988**	3.050	4.980**	−3.285	−3.381
Change in natural logarithm of life expectancy	−0.009	−0.026*	−0.024	−0.021	−0.015
Change in the natural logarithm of the poverty level	1.417***	1.495***	1.615***	1.623***	1.497***
The change in the natural logarithm of a healthy life year	0.162***	0.100***	0.115***	0.148***	0.163***
Change in the natural logarithm of the average expected years of schooling	−0.069	−0.045	−0.058	−0.009	−0.035
Change in the natural logarithm of the unemployment rate	−0.024***	−0.033***	−0.042***	−0.028***	−0.026***
Change in the natural logarithm of the emigration rate	0.011***	0.012***	0.010***	0.010***	0.012***
Change in the natural logarithm of the severe housing deprivation rate	−0.009**	−0.010**	−0.007	−0.006	−0.007*
Change in the natural logarithm of government spending on public order and safety	−0.018	−0.025	−0.030	−0.022	−0.023
The natural logarithm of economic inequality	2.861	−2.792	−4.483***	0.050	0.277
The quadratic natural logarithm of economic inequality	−0.443*	0.356	0.633***	−0.007	−0.029
N (number of observations)	213	213	213	213	213
Adjusted R^2	0.709	0.699	0.753	0.712	0.688
F criteria (p-value)	<0.001	<0.001	<0.001	<0.001	<0.001
Breaking point	25.34	50.47	34.45	–	–
Actual range of the economic inequality indicator	from 20.9 to 40.8	from 37.2 to 61.6	from 24.3 to 46.8	from 4.49 to 18.07	from 3.03 to 8.32

Source: Authors' calculations.

Note: Gini 1 – Gini coefficient of equivalized disposable income; Gini 2 – Gini coefficient of equivalized disposable income before social transfers (pensions included in social transfers); Gini 3 – Gini coefficient of equivalized disposable income before social transfers (pensions excluded from social transfers); DK – X/I deciles' differentiation coefficient; S80/S20 – income quintile ratio.

* 90% of significance level.

** 95% of significance level.

*** 99% of significance level.

Table 3.28 The impact of economic inequality on socio-economic progress by group of countries, reflecting through the quality-of-life indicators (dependent variable is indicator of healthy life years)

Variables	Coefficient estimates calculated using HCSEM				
	Indicators of economic inequality				
	Gini 1	Gini 2	Gini 3	DK	S80/S20
Dependent variable	Quality of life (logarithmic indicator of healthy life years)				
	Less wealthy countries				
Constant	−3.214	−1.041	−0.948	−1.929	−2.310
Change in natural logarithm of life expectancy	−0.004	0.011	−0.010	0.003	0.038
Change in the natural logarithm of the poverty level	1.248***	1.499***	1.413***	1.328***	0.987***
The change in the natural logarithm of a healthy life year	0.102**	0.079	0.060	0.108***	0.168***
Change in the natural logarithm of the average expected years of schooling	−0.084	−0.111	−0.166	−0.059	−0.026
Change in the natural logarithm of the unemployment rate	−0.028	−0.045***	−0.018	−0.031*	−0.046***
Change in the natural logarithm of the emigration rate	0.018***	0.015***	0.019***	0.016***	0.015***
Change in the natural logarithm of the severe housing deprivation rate	−0.004	0.013	0.007	−0.002	−0.009
Change in the natural logarithm of government spending on public order and safety	−0.063	−0.016	−0.078*	−0.041	0.009
The natural logarithm of economic inequality	1.141	−0.745	−0.424	0.025	0.475
The quadratic natural logarithm of economic inequality	−0.185	0.091	0.049	−0.005	−0.045
Breaking point	21.70	59.00	–	–	2.06
	The difference between rich and less rich countries				
The pseudo-variable of richer countries	0.893	100.877***	−6.413	−11.243*	−3,872
Change in natural logarithm of life expectancy	−0.029	−0.077*	−0.096***	−0.072	−0.085**
Change in the natural logarithm of the poverty level	−0.182	0.192	0.840	0.311	−0.021

(Continued)

Table 3.28 The impact of economic inequality on socio-economic progress by group of countries, reflecting through the quality-of-life indicators (dependent variable is indicator of healthy life years)   *(Continued)*

	Coefficient estimates calculated using HCSEM				
	Indicators of economic inequality				
Variables	Gini 1	Gini 2	Gini 3	DK	S80/S20
The change in the natural logarithm of a healthy life year	0.035	0.018	−0.085	−0.024	−0.046
Change in the natural logarithm of the average expected years of schooling	0.084	0.143	0.200	0.179	0.021
Change in the natural logarithm of the unemployment rate	0.029	0.076***	−0.008	0.039*	0.053**
Change in the natural logarithm of the emigration rate	−0.008	0.017**	−0.029***	−0.018***	−0.011
Change in the natural logarithm of the severe housing deprivation rate	−0.007	−0.013	−0.012	−0.007	−0.002
Change in the natural logarithm of government spending on public order and safety	0.008	−0.175***	0.065	0.008	−0.058
The natural logarithm of economic inequality	−0.127	−51.975***	1.581	3.010***	1.518
The quadratic natural logarithm of economic inequality	0.027	6.640***	−0.155	−0.220***	−0.117
Breaking point	24.56	50.19	–	8.48	4.82
Actual range of the economic inequality indicator	from 20.9 to 40.8	from 37.2 to 61.6	from 24.3 to 46.8	from 4.49 to 18.07	from 3.03 to 8.32
N (number of observations)	213	213	213	213	213
Adjusted R^2	0.844	0.856	0.758	0.725	0.702
F criteria (p-value)	<0.001	<0.001	<0.001	<0.001	<0.001

Source: Authors' calculations.

Note: Gini 1 – Gini coefficient of equivalized disposable income; Gini 2 – Gini coefficient of equivalized disposable income before social transfers (pensions included in social transfers); Gini 3 – Gini coefficient of equivalized disposable income before social transfers (pensions excluded from social transfers); DK – X/I deciles' differentiation coefficient; S80/S20 – income quintile ratio.

* 90% of significance level.

** 95% of significance level.

*** 99% of significance level.

expressed by Gini 2 coefficient shows an uneven distribution of income for the population without pensions and other social benefits, the negative impact of increase of income inequality on quality of life in terms of healthy life years can be explained.

By assessing economic inequality with X/I deciles' differentiation coefficient, in richer countries the increase in economic inequality has a positive marginal effect on quality of life and the excessive economic inequality is reached than further increase in economic inequality is associated with deteriorating quality of life when X/I deciles' differentiation coefficient reaches 8.48 (see Table 3.28). Thus, economic inequality has a negative impact on quality of life in Italy (X/I − 12.6), Spain (X/I − 11.0), Luxembourg (X/I − 10.3), the United Kingdom (X/I − 10), and Germany (X/I − 9.9).

The impact of economic inequality on the level of emigration. When measuring quality of life by the emigration rate and assessing the impact of economic inequality on quality of life adjusted by the HCSEM, all indicators of economic inequality are statistically significant – the Gini coefficient of equivalized disposable income (Gini 1); the Gini coefficient of equivalized disposable income before social transfers, in the cases of pensions included in social transfers (Gini 2); the Gini coefficient of equivalized disposable income before social transfers, where pensions are eliminated from social transfers (Gini 3); X/I deciles' coefficient of differentiation; and S80/S20 income quintile ratio. As shown in Table 3.29, economic inequality has a positive impact on emigration, i.e., promoting an increase in emigration rates.

Table 3.29 The impact of economic inequality on socio-economic progress, reflecting through the quality-of-life indicators (dependent variable logarithmic emigration rate)

	Coefficient estimates calculated using HCSEM				
	Indicators of economic inequality				
Variables	*Gini 1*	*Gini 2*	*Gini 3*	*DK*	*S80/S20*
Dependent variable	Quality of life (logarithmic emigration rate)				
Constant	−248.29***	−727.05***	−382.21***	−216.45***	−264.01***
Change in natural logarithm of life expectancy	0.392*	−0.064	−0.970***	−0.012	0.443*
Change in the natural logarithm of the poverty level	3.500	16.459***	24.959***	10.631***	1.257
The change in the natural logarithm of a healthy life year	−0.903**	−0.982***	−1.503***	−0.686**	−1.457***

(Continued)

Table 3.29 The impact of economic inequality on socio-economic progress, reflecting through the quality-of-life indicators (dependent variable logarithmic emigration rate)   *(Continued)*

Variables	Coefficient estimates calculated using HCSEM				
	Indicators of economic inequality				
	Gini 1	Gini 2	Gini 3	DK	S80/S20
Change in the natural logarithm of the average expected years of schooling	5.499***	1.370**	3.084***	5.024***	4.743***
Change in the natural logarithm of the unemployment rate	−0.185	−0.448***	−0.540***	−0.561***	−0.194
Change in the natural logarithm of the emigration rate	3.086***	0.282	1.068	3.789***	4.173***
Change in the natural logarithm of the severe housing deprivation rate	0.050	−0.027	−0.075	−0.088**	−0.027
Change in the natural logarithm of government spending on public order and safety	1.989***	2.799***	2.136***	2.064***	2.076***
The natural logarithm of economic inequality	121.361***	330.304***	151.684***	43.498***	73.570***
The quadratic natural logarithm of economic inequality	−17.145***	−41.025***	−20.148***	−3.070***	−5.647***
N (number of observations)	213	213	213	213	213
Adjusted R^2	0.759	0.769	0.768	0.753	0.726
F criteria (p-value)	<0.001	<0.001	<0.001	<0.001	<0.001
Breaking point	34.44	56.02	43.13	11.93	6.74
Actual range of the economic inequality indicator	from 20.9 to 40.8	from 37.2 to 61.6	from 24.3 to 46.8	from 4.49 to 18.07	from 3.03 to 8.32

Source: Authors' calculations.

Note: Gini 1 – Gini coefficient of equivalized disposable income; Gini 2 – Gini coefficient of equivalized disposable income before social transfers (pensions included in social transfers); Gini 3 – Gini coefficient of equivalized disposable income before social transfers (pensions excluded from social transfers); DK – X/I deciles' differentiation coefficient; S80/S20 – income quintile ratio.

★ 90% of significance level.

★★ 95% of significance level.

★★★ 99% of significance level.

Comparing the impact of economic inequality on emigration between countries having higher living standards and countries having lower living standards in the EU–28, the Chow test was applied. The results showed that in a less wealthy group of countries, only one indicator of economic inequality (Gini coefficient of equivalized disposable income before social transfers when pensions are included in social transfers [Gini 2]) is statistically significant, while in a richer group of countries, all indicators of economic inequality used in the calculations are statistically significant (see Table 3.30). These results justify that economic inequality influence the increase in emigration rates. It is worth noting that the analysis of the impact of economic inequality on the level of emigration also highlights a certain breaking point in the change in the direction of impact between the variables in question, i.e., when a certain point is reached, the level of emigration begins to decrease. The statistical significance of economic inequality in the group of richer countries justifies that economic growth does not address the problem of emigration in principle, as, in the face of severe economic inequality, people are less satisfied with their lives. If they are less satisfied with their lives, they have less optimism and are less connected to the particular country (Telešienė, 2017). Therefore, according to A. Telešienė (2017), it is very important to strengthen the "anchors" who keep people in their homeland in order to reduce emigration. Income is a very important criterion, but there are other reasons that deter people. The main "anchors" able to deter these people from emigration – family, strong community, dignified work, and quality housing.

To sum up, the quality of life factors in the EU–28 are influenced by indicators of economic inequality, along with a whole complex of factors: median income, average life expectancy, population at risk of poverty or social exclusion, years of healthy life, average years of schooling, unemployment rate, emigration rate, severe housing deprivation rate, government spending on public order, and safety.

The impact of economic inequality on quality of life is characterized by a breaking point, followed by a change in the trend(s) of economic inequality impact on quality of life. When assessing the impact of economic inequality on quality of life, the most pronounced negative impact of excessive inequality, where the marginal effect of economic inequality on quality of life becomes negative and the further increase in economic inequality is linked to a deterioration in quality of life, is obtained by applying Gini coefficient of equivalized disposable income (Gini 1), the X/I deciles' differentiation coefficient, and the S80/S20 income quintile ratio. The results show that the increase in economic inequality has a positive marginal effect on the quality of life of EU countries until the following levels are reached: 29.24%, 7.92, and 4.64, respectively. After this breaking point, the marginal effect of economic inequality on quality of life becomes negative in EU–28 countries.

The impact of economic inequality on quality of life is different for different groups of countries according to living standards. In richer countries,

Table 3.30 The impact of economic inequality on socio-economic progress by group of countries, reflecting through the quality-of-life indicators (dependent variable is rate of emigration)

	Coefficient estimates calculated using HCSEM				
	Indicators of economic inequality				
Variables	Gini 1	Gini 2	Gini 3	DK	S80/S20
Dependent variable	Quality of life (logarithmic emigration rate) Less wealthy countries				
Constant	−151.75***	−406.24***	−162.34***	−60.109	−84.879
Change in natural logarithm of life expectancy	−1.220**	−1.318***	−1.530***	−1.432***	−1.374***
Change in the natural logarithm of the poverty level	7.062	11.281**	8.727**	11.654**	9.719**
The change in the natural logarithm of a healthy life year	1.148*	1.196**	0.710	0.974	0.979
Change in the natural logarithm of the average expected years of schooling	11.044***	9.692***	10.471***	9.224***	10.103***
Change in the natural logarithm of the unemployment rate	−0.446*	−0.458**	−0.320	−0.422	−0.396*
Change in the natural logarithm of the emigration rate	5.777***	3.265**	6.017***	6.411***	6.751***
Change in the natural logarithm of the severe housing deprivation rate	−0.077	−0.288**	−0.139	0.010	−0.022
Change in the natural logarithm of govern ment spending on public order and safety	0.993	1.742***	1.427**	1.433*	1.412**
The natural logarithm of economic inequality	48.882	169.303***	50.206	−7.548	1.094
The quadratic natural logarithm of economic inequality	−7.054	−21.363***	−6.824	0.617	−0.017
Breaking point	31.98	52.59	39.60	4.52	–
	The difference between rich and less rich countries				
The pseudo-variable of richer countries	−247.427**	−617.38***	−419.32***	−273.806***	611.555***
Change in natural logarithm of life expectancy	0.909	0.468	0.049	1.126**	1.277**
Change in the natural logarithm of the poverty level	−4.872	10.043	24.020	−1.699	−9.565

(Continued)

Table 3.30 The impact of economic inequality on socio-economic progress by group of countries, reflecting through the quality-of-life indicators (dependent variable is rate of emigration) *(Continued)*

	Coefficient estimates calculated using HCSEM				
	Indicators of economic inequality				
Variables	Gini 1	Gini 2	Gini 3	DK	S80/S20
The change in the natural logarithm of a healthy life year	−4.007***	−1.972***	−3.700***	−2.558***	−4.740***
Change in the natural logarithm of the average expected years of schooling	−5.263**	−9.637***	−8.131***	−3.713	−3.681*
Change in the natural logarithm of the unemployment rate	0.460	−0.293	−0.933***	0.352	0.827***
Change in the natural logarithm of the emigration rate	−4.829***	−4.498***	−8.645***	−6.859***	−7.089***
Change in the natural logarithm of the severe housing deprivation rate	0.067	0.161	0.244	−0.106	0.017
Change in the natural logarithm of government spending on public order and safety	1.477**	1.280**	2.130***	1.285*	1.307*
The natural logarithm of economic inequality	173.663***	309.558***	202.345***	92.684***	220.269***
The quadratic natural logarithm of economic inequality	−24.504***	−38.504***	−26.444***	−6.750***	−17.426***
Breaking point	33.99	54.57	44.51	10.34	5.70
Actual range of the economic inequality indicator	from 20.9 to 40.8	from 37.2 to 61.6	from 24.3 to 46.8	from 4.49 to 18.07	from 3.03 to 8.32
N (number of observations)	213	213	213	213	213
Adjusted R^2	0.878	0.828	0.857	0.960	0.823
F criteria (p-value)	<0.001	<0.001	<0.001	<0.001	<0.001

Source: Authors' calculations.

Note: Gini 1 – Gini coefficient of equivalized disposable income; Gini 2 – Gini coefficient of equivalized disposable income before social transfers (pensions included in social transfers); Gini 3 – Gini coefficient of equivalized disposable income before social transfers (pensions excluded from social transfers); DK – X/I deciles' differentiation coefficient; S80/S20 – income quintile ratio.

* 90% of significance level.

** 95% of significance level.

*** 99% of significance level.

all indicators of economic inequality are statistically significant. In richer countries, the increase in economic inequality has a positive impact on quality of life and later the marginal effect of economic inequality on quality of life becomes negative when Gini 1 coefficient reaches the level of 28.87%, Gini 2 – 52.17%, Gini 3 – 41.00, X/I – 8.18, and S80/S20 – 4.90. In a less wealthy group of countries, just one economic inequality indicator is statistically significant – Gini 2 coefficient. The results show that in a less wealthy group of countries, the increase in economic inequality has a positive marginal effect on quality of life and, just when Gini 2 coefficient reaches 40.86%, the marginal effect of income inequality on quality of life becomes a negative.

According to the research, excessive inequality that negatively affects the countries' quality of life is more pronounced in more than half of the EU countries: Bulgaria, Romania, Hungary, Lithuania, Latvia, Estonia, Poland, Italy, Spain, Portugal, Croatia, Greece, Ireland, the United Kingdom, Germany, and Luxembourg.

The assessment of economic inequality by the Gini coefficient of equivalized disposable income (Gini 1) and the measuring quality of life based on indicator of healthy life years shows that the positive marginal effect of economic inequality on quality of life in EU countries is at a breaking point (25.34%). After reaching this breaking point, economic inequality become excessive as the marginal effect of economic inequality on quality of life becomes negative and the further increase in economic inequality is linked to a deteriorating quality of life. On the basis of this result, economic inequality has a negative impact on the quality of life of almost all EU countries, as in 2018, in almost all EU countries, Gini 1 was more than 25.34% (with the exception of the Czech Republic, Slovenia, and Slovakia).

When measuring quality of life by the indicator of emigration rate and assessing the impact of economic inequality on quality of life, all indicators of economic inequality used in the calculations are statistically significant. The results confirm that economic inequality is driving an increase in emigration rates.

Comparing the impact of economic inequality on emigration rate between countries having the higher living standards and countries having the lower living standards in the EU-28 and applying the Chow test results showed that in a richer group of countries, all indicators of economic inequality used in the calculations are statistically significant. This justifies idea that economic inequality is influencing the increase in emigration rates. It is worth noting that the analysis of the impact of economic inequality on the emigration rate also highlights a certain breaking point in the change in the direction of impact between the variables in question. When a certain point is reached, the rate of emigration begins to decrease. The statistical significance of economic inequality impacts in the group of richer countries justifies the idea that economic growth and the country's standard of living does not address the problem of emigration in principle.

Therefore, in order to reduce emigration, it is very important to strengthen the other anchors that keep people in their homeland – dignified work and income, family, strong community, and quality housing.

3.6 Paradoxical impact of increasing inequality on the environment

Assessing sustainability by measuring forest area (as the share of the total area of the country, %) and assessing the impact of economic inequality on impact on environmental sustainability using the LSM, the hypothesis (H0) is tested that not all indicators have an impact on sustainability. Based on the results (see Table 3.31), the complex significance criterion F (*p*-value) of the indicators is low, demonstrating a high reliability of the calculations and results obtained; therefore, a hypothesis is rejected. Therefore, the distinguished indicators (together with indicators of economic inequality) have an impact on sustainability.

Based on the calculations made (see Table 3.31), statistically significant indicators of economic inequality are Gini coefficient of equivalized disposable income (Gini 1), the X/I deciles' differentiation coefficient, and

Table 3.31 The impact of economic inequality on socio-economic progress, reflecting through sustainability indicators (dependent variable is logarithmic forest area indicator)

	Coefficient estimates calculated using SSA				
	Indicators of economic inequality				
Variables	*Gini 1*	*Gini 2*	*Gini 3*	*DK*	*S80/S20*
Dependent variable	Sustainability (logarithmic indicator of forest areas)				
Constant	269.967***	166.353*	9.502	88.692**	180.728***
Change in the natural logarithm of government expenditure on the environmental protection	−0.440***	−0.432***	−0.458***	−0.472***	−0.478***
Change in the natural logarithm of fossil fuel energy consumption	−0.622***	−0.795***	0.842***	−0.813***	−0.794***
Change in natural logarithm of GDP per capita	3.615***	2.963***	3.165***	2.872***	2.983***
Change in natural logarithm of consumption level	−3.797***	−3.136***	−3.386***	−3.069***	−3.156***

(Continued)

Table 3.31 The impact of economic inequality on socio-economic progress, reflecting through sustainability indicators (dependent variable is logarithmic forest area indicator)   *(Continued)*

Variables	Coefficient estimates calculated using SSA				
	Indicators of economic inequality				
	Gini 1	Gini 2	Gini 3	DK	S80/S20
Change in the natural logarithm of investments	−0.095	0.288	−0.181	−0.045	−0.161
Change in natural logarithm of trade (imports, exports)	−1.168***	−1.044***	−1.050***	−0.968***	−0.982***
Change in the natural logarithm of the corruption perception index	−0.436	−0.604	−0.537	−0.632	−0.628
The natural logarithm of economic inequality	−150.84***	−77.406	5.131	−21.842**	−53.366***
The quadratic natural logarithm of economic inequality	22.169***	9.798	−0.982	1.607**	4.282***
N (number of observations)	104	104	104	104	104
Adjusted R^2	0.545	0.440	0.474	0.454	0.472
F criteria (p-value)	<0.001	<0.001	<0.001	<0.001	<0.001
The value of test p to check for autocorrelation	0.000	0.000	0.000	0.000	0.000
Value of test p to check heteroskedasticity	0.000	0.000	0.000	0.000	0.000
Breaking point	30.03	51.94	−	8.93	5.08
Actual range of the economic inequality indicator	from 20.9 to 40.8	from 37.2 to 61.6	from 24.3 to 46.8	from 4.49 to 18.07	from 3.03 to 8.32

Source: Authors' calculations.

Note: Gini 1 – Gini coefficient of equivalized disposable income; Gini 2 – Gini coefficient of equivalized disposable income before social transfers (pensions included in social transfers); Gini 3 – Gini coefficient of equivalized disposable income before social transfers (pensions excluded from social transfers); DK – X/I deciles' differentiation coefficient; S80/S20 – income quintile ratio.

* 90% of significance level.

** 95% of significance level.

*** 99% of significance level.

the S80/S20 income quintile ratio. The results confirm that economic inequality has a negative impact on sustainability, i.e., reducing the area of
forests in the country. It should be noted that when analyzing the impact of
economic inequality on sustainability, a certain breaking point is evident
when the direction of impact between the variables in question changes.
When economic inequality is measured by Gini coefficient of equivalized
disposable income (Gini 1), the breaking point is at the 33.03% threshold
(see Figure 3.22). This can be explained by assuming that the greater the
economic inequality, the less negative is the impact on sustainability. Since
high economic inequality mean a large number of poor people compared
to a small number of rich people, the negative impact on the planet and
its sustainability is much less. In countries with high economic inequality,
the negative impact on the sustainability can be largely influenced by the
richest members of society. Meanwhile, in countries where there is no very
large income gap and economic inequality is not high, the intensive use of
resources, active industrial development and high public consumption can be
noticed, providing for the significant negative impact on sustainability.

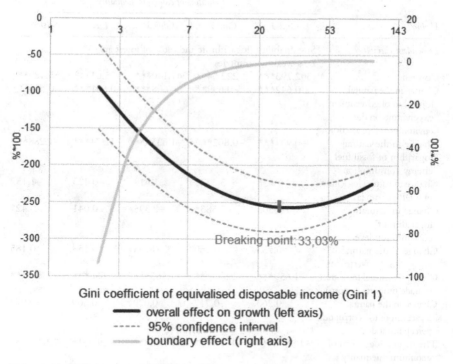

Figure 3.22 Relative slope of the impact of economic inequality (according to Gini 1) on
socio-economic progress, reflected through sustainability indicators (logarithmic forest area indicator).

Source: Authors' calculations.

The Chow test applied to assess the impact of economic inequality on environmental sustainability in different groups of countries: richer countries having the higher standard of life (where GDP per capita in PPS is more than €24750) and in less wealthy countries (where GDP per capita under PPS is less than €24750), revealed statistically significant indicators of economic inequality in the less wealthy group of countries. Meanwhile, in the group of richer countries, only one indicator of economic inequality is statistically significant: the Gini coefficient of equivalized disposable income before social transfers and pensions eliminated from social transfers (Gini 3) (see Table 3.32). In all cases examined, economic inequality has a negative impact on sustainability, especially in the less wealthy group of countries.

Table 3.32 The impact of economic inequality on socio-economic progress by group of countries, reflecting through sustainability indicators (the dependent variable is logarithmic forest area indicator)

	Coefficient estimates calculated using SSA				
	Indicators of economic inequality				
Variables	*Gini 1*	*Gini 2*	*Gini 3*	*DK*	*S80/S20*
Dependent variable	Sustainability (logarithmic indicator of forest areas)				
	Less wealthy countries				
Constant	302.793***	222.649*	347.140***	125.017**	255.429***
Change in the natural logarithm of government expenditure on the environmental protection	−0.642***	−0.699**	−0.608***	−0.661***	−0.654***
Change in the natural logarithm of fossil fuel energy consumption	−0.931***	−0.862***	−1.031***	−1.274***	−1.288***
Change in natural logarithm of GDP per capita	1.638	3.343*	1.619	−0.423	−1.152
Change in natural logarithm of consumption level	−2.505	−3.583*	−2.395	−0.041	0.421
Change in the natural logarithm of investments	−0.753	0.160	−0.795	0.158	−0.185
Change in natural logarithm of trade (imports, exports)	−1.659***	−1.548***	−1.565***	−1.658***	−1.649***
Change in the natural logarithm of the corruption perception index	0.263	−0.717	0.079	−0.669	−0.073
The natural logarithm of economic inequality	−162.39***	−103.772*	−180.68***	−29.031*	−72.534***
The quadratic natural logarithm of economic inequality	23.575***	13.117*	25.202***	2.060*	5.685***
Breaking point	31.32	52.23	36.04	11.48	5.90

(Continued)

Table 3.32 The impact of economic inequality on socio-economic progress by group of countries, reflecting through sustainability indicators (the dependent variable is logarithmic forest area indicator) *(Continued)*

Variables	Coefficient estimates calculated using SSA				
	Indicators of economic inequality				
	Gini 1	Gini 2	Gini 3	DK	S80/S20
	The difference between rich and less rich countries				
The pseudo-variable of richer countries	86.123	−12.468	−343.062**	24.069	54.443
Change in the natural logarithm of government expenditure on the environmental protection	0.432*	0.520	0.359	0.487*	0.465*
Change in the natural logarithm of fossil fuel energy consumption	0.616	0.336	0.556	0.962**	0.931**
Change in natural logarithm of GDP per capita	0.486	−0.808	0.887	2.359	2.952
Change in natural logarithm of consumption level	2.224	2.450	1.112	−0.026	−0.279
Change in the natural logarithm of investments	4.003***	2.369*	2.914**	3.086***	3.491***
Change in natural logarithm of trade (imports, exports)	0.819*	0.461	0.592	0.807*	0.870**
Change in the natural logarithm of the corruption perception index	−1.849**	−0.574	−1.272	−0.586	−1.467
The natural logarithm of economic inequality	−77.145	−5.254	177.591**	−20.503	−33.608
The quadratic natural logarithm of economic inequality	11.994	0.567	−25.046**	1.659	2.968
Breaking point	29.00	53.72	–	7.81	4.61
Actual range of the economic inequality indicator	from 20.9 to 40.8	from 37.2 to 61.6	from 24.3 to 46.8	from 4.49 to 18.07	from 3.03 to 8.32
N (number of observations)	104	104	104	104	104
Adjusted R^2	0.699	0.535	0.621	0.634	0.663
F criteria (p-value)	<0.001	<0.001	<0.001	<0.001	<0.001

Source: Authors' calculations.

Note: Gini 1 – Gini coefficient of equivalized disposable income; Gini 2 – Gini coefficient of equivalized disposable income before social transfers (pensions included in social transfers); Gini 3 – Gini coefficient of equivalized disposable income before social transfers (pensions excluded from social transfers); DK – X/I deciles' differentiation coefficient; S80/S20 – income quintile ratio.

* 90% of significance level.

** 95% of significance level.

*** 99% of significance level.

The tests for reliability of the LSM model were carried out. First, the Wooldridge test for autocorrelation, when the hypothesis (H0) was tested, that the model is characterized by autocorrelation and the obtained test p-values <0.001 (see Table 3.32) indicate that there is autocorrelation in the model. Secondly, the White test was applied to determine heteroskedicity when testing the hypothesis (H0) that the model is characterized by heteroskedicity and the test p-value <0.001 (see Table 3.31) confirms the existence of heteroskedicity. Therefore, in order to provide a more reliable model for assessing the impact of economic inequality on sustainability, as well as addressing the problems of autocorrelation and heteroskedicity, the model was recalculated by the method of the HCSEM.

In assessing the impact of economic inequality on sustainability under the HCSEM, almost all indicators of economic inequality (except for the Gini coefficient of equivalized disposable income before social transfers, where pensions are eliminated from social transfers [Gini 3]) are statistically significant justifying that economic inequality adversely affect sustainability until the certain breaking points: Gini 1 – 30.92%, Gini 2 – 52.52%, X/I deciles' coefficient of differentiation – 10.96, S80/S20 income quintile ratio – 5.52, and then change direction, and the marginal effect of economic inequality on sustainability becomes positive (see Table 3.33).

By dividing the EU-28 countries into countries having higher living standards and countries having lower living standards and following the Chow test, the results showed that in countries with lower living standards, all indicators of economic inequality are statistically significant – the Gini coefficient of equivalized disposable income (Gini 1); the Gini coefficient of equivalized disposable income before social transfers, in the case of pensions included in social transfers (Gini 2); the Gini coefficient of equivalized disposable income before social transfers, where pensions are eliminated from social transfers (Gini 3); the X/I deciles' coefficient of differentiation; and the 80/S20 income quintile ratio (see Table 3.34); and thus, the marginal effect of economic inequality on sustainability in all cases is negative. However, when a certain level of economic inequality or turning point is reached, the marginal effect of economic inequality on sustainability becomes positive. In a group of less wealthy countries, the breaking point is recorded at 31.99%, measured by Gini 1, at 55.16% – by Gini 2; and at 37.51% – by Gini 3, 10.58 for X/I deciles' differentiation coefficient, and 5.61 for S80/S20 quintiles income ratio (see Table 3.34). This again highlights the findings of research that the greater the economic inequality, the less negative is the impact on sustainability.

In a group of more wealthy countries, two indicators of economic inequality are statistically significant: the Gini coefficient of equivalized disposable income (Gini 1) and the Gini coefficient of equivalized disposable income before social transfers, where pensions are eliminated from social transfers (Gini 3). The results show that in this group of countries, the increase in economic inequality has a positive marginal effect on sustainability and,

Table 3.33 The impact of economic inequality on socio-economic progress, reflecting through sustainability indicators (the dependent variable is logarithmic forest area indicator)

Variables	Coefficient estimates calculated using HCSEM				
	Indicators of economic inequality				
	Gini 1	Gini 2	Gini 3	DK	S80/S20
Dependent variable	Sustainability (logarithmic indicator of forest areas)				
Constant	205.359***	311.670***	87.868***	78.501***	162.634***
Change in the natural logarithm of government expenditure on the environmental protection	−0.337***	−0.230***	−0.373***	−0.249***	−0.277***
Change in the natural logarithm of fossil fuel energy consumption	−0.899***	−1.071***	−1.031***	−0.892***	−0.851***
Change in natural logarithm of GDP per capita	1.090	0.782	0.627	0.390	0.722
Change in natural logarithm of consumption level	−1.335*	−1.136*	−0.899	−0.673	−1.003
Change in the natural logarithm of investments	−0.336	0.314	−0.376*	−0.204	−0.261
Change in natural logarithm of trade (imports, exports)	−0.731***	−0.980***	−0.747***	−0.812***	−0.767***
Change in the natural logarithm of the corruption perception index	−0.598**	−0.129	−0.202	−0.637***	−0.580***
The natural logarithm of economic inequality	−110.62***	−149.87***	−38.339	17.765***	−46.534***
The quadratic natural logarithm of economic inequality	16.118***	18.918***	5.047	1.269***	3.685***
N (number of observations)	104	104	104	104	104
Adjusted R^2	0.842	0.879	0.916	0.832	0.825
F criteria (p-value)	<0.001	<0.001	<0.001	<0.001	<0.001
Breaking point	30.92	52.52	44.61	10.96	5.52
Actual range of the economic inequality indicator	from 20.9 to 40.8	from 37.2 to 61.6	from 24.3 to 46.8	from 4.49 to 18.07	from 3.03 to 8.32

Source: Authors' calculations.

Note: Gini 1 – Gini coefficient of equivalized disposable income; Gini 2 – Gini coefficient of equivalized disposable income before social transfers (pensions included in social transfers); Gini 3 – Gini coefficient of equivalized disposable income before social transfers (pensions excluded from social transfers); DK – X/I deciles' differentiation coefficient; S80/S20 – income quintile ratio

* 90% of significance level.

** 95% of significance level.

*** 99% of significance level.

Table 3.34 The impact of economic inequality on socio-economic progress by group of countries, reflecting through sustainability indicators (dependent variable is logarithmic forest area indicator)

| | Coefficient estimates calculated using HCSEM | | | | |
| | Indicators of economic inequality | | | | |
Variables	Gini 1	Gini 2	Gini 3	DK	S80/S20
Dependent variable	Sustainability (logarithmic indicator of forest areas)				
	Less wealthy countries				
Constant	221.188***	164.933***	250.180***	146.249***	181.237***
Change in the natural logarithm of government expenditure on the environmental protection	−0.211**	−0.025	−0.143	−0.206	−0.010
Change in the natural logarithm of fossil fuel energy consumption	−0.766***	−0.901***	−1.069***	−1.167***	−0.724***
Change in natural logarithm of GDP per capita	0.043	−0.377	−1.060	−1.497	−2.531***
Change in natural logarithm of consumption level	−0.679	0.167	0.344	0.921	2.196***
Change in the natural logarithm of investments	−0.475*	−0.050	−0.448	−0.173	0.150
Change in natural logarithm of trade (imports, exports)	−1.135***	−0.571**	−1.151***	−1.194***	−0.507**
Change in the natural logarithm of the corruption perception index	0.502***	0.161	0.556**	0.381	0.468*
The natural logarithm of economic inequality	−117.90***	−76.583**	−127.48***	−36.709***	−53.974***
The quadratic natural logarithm of economic inequality	17.011***	9.549**	17.585***	2.636***	4.263***
Breaking point	31.99	55.16	37.51	10.58	5.61
	The difference between rich and less rich countries				
The pseudo-variable of richer countries	−133.288**	45.809	−406.79***	−32.534	7.909
Change in the natural logarithm of government expenditure on the environmental protection	−0.253*	−0.481**	−0.464***	−0.116	−0.385**

(Continued)

Table 3.34 The impact of economic inequality on socio-economic progress by group of countries, reflecting through sustainability indicators (dependent variable is logarithmic forest area indicator)   *(Continued)*

	Coefficient estimates calculated using HCSEM				
	Indicators of economic inequality				
Variables	Gini 1	Gini 2	Gini 3	DK	S80/S20
Change in the natural logarithm of fossil fuel energy consumption	−0.117	−0.205	0.162	0.813***	−0.018
Change in natural logarithm of GDP per capita	−0.204	0.070	1.059	2.078*	2.434***
Change in natural logarithm of consumption level	1.971*	1.013	−0.039	0.572	−0.753
Change in the natural logarithm of investments	1.904***	1.007	0.826	2.758***	1.745***
Change in natural logarithm of trade (imports, exports)	1.313***	0.884***	1.508***	0.828**	0.531*
Change in the natural logarithm of the corruption perception index	−2.411***	−2.036***	−1.781***	−1.649***	−1.945***
The natural logarithm of economic inequality	65.069*	−28.342	219.526***	−2.865	−10.269
The quadratic natural logarithm of economic inequality	−9.203	3.685	−30.573***	0.352	1.002
Breaking point	29.47	52.69	34.59	7.52	4.47
Actual range of the economic inequality indicator	from 20.9 to 40.8	from 37.2 to 61.6	from 24.3 to 46.8	from 4.49 to 18.07	from 3.03 to 8.32
N (number of observations)	104	104	104	104	104
Adjusted R^2	0.924	0.953	0.909	0.910	0.931
F criteria (p-value)	<0.001	<0.001	<0.001	<0.001	<0.001

Source: Authors' calculations.

Note: Gini 1 – Gini coefficient of equivalized disposable income; Gini 2 – Gini coefficient of equivalized disposable income before social transfers (pensions included in social transfers); Gini 3 – Gini coefficient of equivalized disposable income before social transfers (pensions excluded from social transfers); DK – X/I deciles' differentiation coefficient; S80/S20 – income quintile ratio.

* 90% of significance level.

** 95% of significance level.

*** 99% of significance level.

with Gini 1 reaching 29.47% and Gini 3 − 34.59%, the marginal effect of economic inequality becomes negative. According to 2018 statistics, economic inequality had a negative impact on environmental sustainability in the United Kingdom (Gini 1 − 33.5%, Gini 3 − 40.4%), Italy (Gini 1 − 33.4%, Gini 3 − 35.7%), Spain (Gini 1 − 33.2%, Gini 3 − 37%), Luxembourg (Gini 1 − 33.2%, Gini 3 − 38.1%), Germany (Gini 1 − 31.1%, Gini 3 − 36.6%), Ireland (Gini 3 − 39.3%), Danish (Gini 3 − 36%), Sweden (Gini 3 − 35.7%), and France (Gini 3 − 34.9%).

In summary, the results of the assessment of economic inequality's impact on sustainability suggest that economic inequality has a negative impact on sustainability. A more detailed analysis of the impact of economic inequality on sustainability highlights a certain breaking point in the change in the direction of impact between the variables in question, i.e., when a certain breaking point is reached, the marginal effect of economic inequality on sustainability changes from negative to positive. This can be explained by assuming that the greater is the economic inequality, the less negative is the impact on sustainability since due to high economic inequality a large number of poor people compared to a small number of rich people exist. This means restricted consumption of high number of pour people providing smaller impact on environmental sustainability of the country. Meanwhile, in countries where there is no very large income gap − economic inequality is not high, the intensive resource use, active industrial development and high public consumption can be noticed, and therefore, in such countries, the negative impact on sustainability is significant.

The main trend is that marginal effects of economic inequality on sustainability are negative, but at a certain level − a breaking point − the marginal effect of economic inequality on sustainability becomes positive, especially in a group of less wealthy countries. In a group of less wealthy countries, the following breaking points are recorded: 31.99%, measured by Gini 1, 55.16% measured by Gini 2, and 37.51% measured by Gini 3. In a group of more wealthy countries, the marginal effect of economic inequality on sustainability is first and foremost positive; however, with Gini 1 coefficient reaching 29.47% and Gini 3 coefficient − 34.59%, the marginal effect of economic inequality on sustainability becomes negative. It is therefore possible to suggest that based on 2018 data, economic inequality had a negative impact on the sustainability of Ireland, Denmark, Italy, Spain, the United Kingdom, Luxembourg, France, Sweden and Germany.

3.7 Generalized impact of excessive inequality on socio-economic progress

Socio-economic progress is a complex concept and a complex process aimed at ensuring prosperity for current and future generations. The integrity of economic growth, quality of life, and sustainability is a precondition for ensuring socio-economic progress. In order to assess the impact of economic inequality on socio-economic progress in the EU-28, factors influencing economic inequality were identified and the interaction of the components of economic inequality and indicators of socio-economic progress (growth, quality of life, and sustainability) was analyzed. The research provides the following results in assessing the impact of economic inequality on socio-economic progress in EU countries.

- *The stronger impact of economic inequality on economic growth is occurring in the longer term.*

  1   In analyzing the *impact of economic inequality on the economic growth rate of the current year and for the two-year average growth rate*, the statistically significant economic inequality indicator, i.e., Gini coefficient of equivalized disposable income before social transfers when pensions were eliminated from social transfers (Gini 3) was revealed in the group of richer countries. In the latter group of countries, the increase in economic inequality has a positive marginal effect on economic growth, with Gini 3 coefficient being no more than 34.03% (for current year) and 34.98% (for second year) and after reaching this point, the economic inequality becomes excessive. This means that the marginal effect of economic inequality on economic growth is becoming negative and the continued increase in economic inequality slows down economic growth.

  2   Three indicators of economic inequality were statistically significant for the *three-year average growth rate* in the wealthier countries group:

    1   Gini coefficient of equivalized disposable income (Gini 1),
    2   Gini coefficient of equivalized disposable income before social transfers when pensions are included in social transfers (Gini 2), and
    3   Gini coefficient of equivalized disposable income before social transfers, when pensions are eliminated from social transfers (Gini 3).

The results confirmed that the marginal effect of economic inequality on economic growth is positive to the breaking points: for Gini 1 – 29.63%, Gini 2 – 53.06%, and Gini 3 – 36.45%. After that, the economic inequality in these breaking points become excessive and the marginal effect of economic inequality on economic growth becomes negative, and the further increase in economic inequality slows down economic growth.

  3   Although the *assessment of the impact of economic inequality on the four- and five-year average economic growth rates* did not reveal a statistically significant indicator of economic inequality, the low complex significance of the criterion F (p-value) of the indicators showed a high reliability of

the results obtained, suggesting that all the identified factors, including economic inequality, have an impact on economic growth all together in the long term.

- *Changes in economic inequality have an uneven impact on socio-economic progress (economic growth, quality of life, environmental sustainability) in groups of countries with different standard of living.*

1   The results of the empirical research show that *excessive inequality, which has a negative impact on economic growth, is more pronounced in more affluent countries.* Economies in less wealthy countries require income growth, even if it exacerbates economic inequality and widens the gap between the richest and poorest members of society. The research shows that economic inequality is excessive and has a negative impact on the economies of richest EU countries: Ireland, Denmark, Estonia, Italy, Spain, the United Kingdom, Luxembourg, France, Finland, Sweden, and Germany.

2   The impact of economic inequality on quality of life is different for different groups of countries in terms of living standards. In richer countries, all indicators of economic inequality are statistically significant:

1   Gini coefficient of equivalized disposable income (Gini 1),
2   Gini coefficient of equivalized disposable income before social transfers when pensions are included in social transfers (Gini 2),
3   Gini coefficient of equivalized disposable income before social transfers, where pensions are eliminated from social transfers (Gini 3),
4   X/I deciles' differentiation coefficient, and
5   S80/S20 income quintile ratio.

In richer countries, the increase in economic inequality has a positive marginal effect on quality of life and economic inequality becomes excessive (the marginal effect of economic inequality on economic growth becomes negative) when Gini 1 reaches the level of 28.87%, Gini 2 – 52.17%, Gini 3 – 41.00, X/I deciles' differentiation ratio – 8.18, and S80/S20 income quintile ratio – 4.90. In a less wealthy group of countries, economic inequality is statistically significant just for one indicators of economic inequality identified – Gini 2 coefficient. The results show that in a less wealthy group of countries, the increase in economic inequality has a positive marginal effect on quality of life and, when Gini 2 reaches 40.86%, the marginal effect of economic inequality on quality of life becomes negative.

According to the research, excessive economic inequality affecting countries' quality of life is more pronounced in more than half of the EU countries: Bulgaria, Romania, Hungary, Lithuania, Latvia, Estonia, Poland, Italy, Spain, Portugal, Croatia, Greece, Ireland, the United Kingdom, Germany, and Luxembourg.

3   Comparing *the impact of economic inequality on emigration in* countries
    with higher and lower living standards in EU-28, the results showed
    that in a richer group of countries, all indicators of economic inequal-
    ity used in the calculations are statistically significant, which justifies
    that economic inequality is affecting the rise in emigration rates. It is
    worth noting that the analysis of the impact of economic inequality on
    the level of emigration also highlights a certain breaking point in the
    change in the direction of impact between the variables in question,
    i.e., when a certain point of economic inequality is reached, the level of
    emigration begins to decrease. The statistical significance of economic
    inequality in the group of richer countries justifies the idea that eco-
    nomic growth (the country's standard of living) does not address the
    problem of emigration in principle.

4   *The results of the assessment of economic* inequality impact on sustainability
    show that economic inequality has a negative impact on sustainability.
    The marginal effects of economic inequality on sustainability are neg-
    ative, but at a certain level – a breaking point – the marginal effect of
    economic inequality on sustainability becomes positive, especially in a
    group of less wealthy countries. In a group of less wealthy countries,
    the breaking point is recorded at Gini 1 coefficient – 31.99%, Gini 2
    coefficient – 55.16%, Gini 3 coefficient – 37.51%, X/I deciles' differen-
    tiation coefficient – 10.58, and S80/S20 income quintile ratio – 5.61. It
    can therefore be argued that economic inequality had a negative impact
    on the sustainability of Bulgaria, Lithuania, Latvia, Romania, Greece,
    and Portugal in 2018.

• *Normal economic inequality affects social economic progress in one direction, and
  when economic inequality become excessive, it has a negative impact on social economic progress (inverted "U" curve).*

1   In addition to the above-mentioned empirical test results that also
    highlight the breaking points, it is possible to state that, when analyzing
    the impact of *economic inequality on quality of life, the latter being measured
    as a median income indicator,* excessive inequality is evident when Gini 1
    coefficient is at – 29.24%, the X/I deciles' coefficient of differentiation –
    7.92, and S80/S20 income quintile ratio – 4.64. In addition, after these
    breaking points, the marginal effect of economic inequality on quality
    of life in the EU-28 is negative. According to research results, excessive
    inequality affecting countries' quality of life is evident in half of the EU
    countries: Bulgaria, Lithuania, Latvia, Romania, the United Kingdom,
    Italy, Spain, Luxembourg, Greece, Portugal, Germany, Estonia, and
    Croatia.

2   Economic inequality assessing by the Gini 1 coefficient on the basis of
    disposable income, and measuring *quality of life as the indicator of healthy
    life years,* shows that the positive marginal effect of economic inequality

on quality of life in EU countries is at a breaking point (25.34%), and economic inequality becomes excessive when the marginal effect of economic inequality on quality of life becomes negative and the further increase in economic inequality is linked to a deteriorating quality of life. On the basis of research results, economic inequality had a negative impact on the quality of life of almost all EU countries, in 2018, as Gini 1 coefficient was more than 25.34% in all EU countries with the exception of the Czech Republic, Slovenia, and Slovakia.

In summary, the following conclusions can be drawn from the empirical research.

- Economic inequality negatively affects the social economic progress of most EU countries. The high level of economic development of countries does not yet guarantee protection against the negative impact of excessive inequality on the country's social economic progress indicators, such as economic growth, quality of life, and environmental sustainability. The results of the empirical research confirm that excessive inequality can have a negative impact on the social progress of both rich countries (e.g., Ireland, Denmark, Estonia, the United Kingdom, Italy, Spain, Luxembourg, France, Finland, Sweden, and Germany) and less wealthy countries (e.g., Bulgaria, Greece, Croatia, Latvia, Poland, Lithuania, Portugal, Romania, and Hungary). Ultimately, those effects and damages may occur in the long term.
- The negative impact of economic inequality on Lithuania's economic growth is most pronounced in the fourth and fifth years of economic growth. In addition, it can be said that the current level of economic inequality in Lithuania is excessive and has a negative impact not only on economic growth, but also on the quality of life and environmental sustainability.
- The negative impact of excessive inequality on social economic progress can be avoided by integrated systemic solutions aiming at reduction of the level of economic inequality. The empirical results of the research confirmed that economic inequality cannot exceed 29–30% by measuring it based on the Gini coefficient of equivalized disposable income in order to avoid the negative impact of excessive inequality on their social economic progress.
- Economic inequality affects the social economic progress of the EU-28 countries – economic growth, quality of life, and sustainability. The impact of economic inequality on the social economic progress of EU countries – economic growth, quality of life, and environmental sustainability – is nonlinear. The impact of economic inequality on economic growth varies from one to another group of EU countries according to living standards.
- Excessive inequality, where the marginal effect of economic inequality on economic growth becomes negative, is more pronounced in the

group of richer countries. For less wealthy countries, it is important to guarantee the growth of general income. Even if that income growth takes place in the richest segment of population, which exacerbates the economic inequality of these countries, the result mitigates the negative impact of economic inequality on the economic growth of poorer countries. The research shows that economic inequality is excessive and has a negative impact on the economies of richest countries such as Ireland, Denmark, Estonia, Italy, Spain, the United Kingdom, Luxembourg, France, Finland, Sweden, and Germany.

- The impact of economic inequality on quality of life, the latter being measured by the median income indicator, is different for groups of countries distinguished according to living standards. The results of the research confirm the existence of excessive inequality, where the marginal effect of economic inequality on quality of life becomes negative and the further increase in economic inequality is linked to a deterioration in quality of life. This excessive inequality was evident in more than half of the EU countries in 2018: Bulgaria, Romania, Hungary, Lithuania, Latvia, Estonia, Poland, Italy, Spain, Portugal, Croatia, Greece, Ireland, the United Kingdom, Germany, and Luxembourg.

- The impact of economic inequality on quality of life in terms of healthy life years shows that economic inequality is transformed into excessive inequality (the marginal effect of economic inequality on quality of life becomes negative) when the breaking point in the Gini coefficient of equivalized disposable income (Gini 1) reaches 25.34%. Based on this result, economic inequality had a negative impact on the quality of life of almost all EU countries in 2018, as Gini 1 exceeded the defined breaking point in almost all EU countries except the Czech Republic, Slovenia, and Slovakia.

- The results of the research confirm that economic inequality contributes to an increase in emigration rates applied as measure of quality of life. By comparing the impact of economic inequality on emigration in EU countries with lower living standards and EU countries with higher living standards, the statistical significance of economic inequality in the richer countries group provides that economic growth (the country's standard of living) does not address the problem of emigration in principle. Therefore, in order to reduce emigration, it is very important to strengthen the "anchors" that keep people in their homeland.

- The results of assessment of economic inequality impact on environmental sustainability suggest that overall economic inequality has a negative impact on the environment. However, a deeper analysis of the impact of economic inequality on environment highlights a certain breaking point in which the direction of impact changes: the marginal effect of economic inequality on environment changes from negative to positive. This can be explained by the paradoxical assumption that the greater is the economic inequality in country, the less negative is the impact on the natural environment. Since severe economic inequality mean a large number of poor people compared to a small number of rich people, the impact on natural environment is much less in this situation. In countries with high economic inequality, the negative impact on the sustainability can be largely influenced by wealthy members of society. Countries that are relatively rich, having relatively more well-off populations and lower income gaps providing for lower economic inequality, are also characterized by intensive resource use, active industrial development, and high public consumption, and therefore with higher negative impact on environmental sustainability than countries having higher economic inequality.

4 GUIDELINES FOR REDUCING EXCESSIVE INEQUALITY

According to the experience of other countries and researchers I. Joumard, M. Pisu, and D. Bloch, benefits reduce income inequality more than taxes. Almost 75% of the reduction in income inequality depends on benefits and the remaining quarter – on taxes.

Based on the results of mentioned research and practical experience of foreign countries (taking into account the population groups most affected by poverty and income and wealth differentiation), the authors emphasize the following main directions and measures to reduce economic inequality (Figure 4.1):

- Introduction of a tax system aimed at a more equitable distribution of income and wealth. Differentiated taxation of income and wealth, application of a progressive tax system, and reduction of the tax burden on the lowest earners – these would be measures to reduce economic inequality;
- Macroeconomic policies focused on full employment. Increasing employment through the inclusion of older people through active labor market policies (not just for the low-skilled and regions' people, whose unemployment rates are significantly higher than high-skilled people living in metropolitan areas). Measures should include education, training, retraining, and competence development, and thus enabling workers to remain in the labor market for as long as possible;
- Implementation of social policies targeting vulnerable and motivated social groups. Ensuring sufficient social support to enable households to meet their basic needs without compromising the ability to seek work. Possible solutions: indexation of pensions, unemployment insurance, social benefit benefits, and the introduction of basic (full or partial) universal salary for vulnerable social groups; and
- Reducing the vulnerability of social groups and strengthening resilience to life difficulties (capacity building and psychological intervention measures, taking into account the most common threats and fears experienced by the population).

DOI: 10.4324/b22984-5

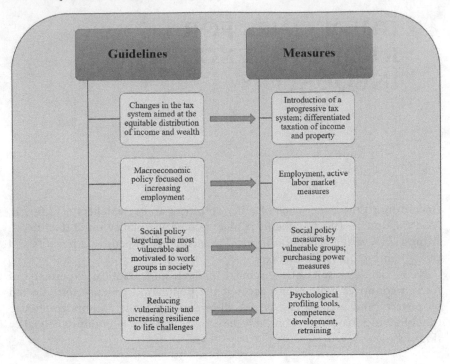

Figure 4.1 Guidelines for reducing the economic inequality.

Source: Compiled by the authors.

4.1 Tax policy measures

Adjustment of the tax system. Countries with higher income inequality tend to redistribute as much income as possible to amortize it. As a result, income taxes are progressive and consumption and property taxes are regressive in many countries.

In Lithuania, most of the state budget revenue consists of consumption taxes (according to the consumption tax, Lithuania is in the third place in the EU), taxation of labor income is on average among the EU countries (15th place in the EU), and capital income taxation is the lowest among the EU countries instead of the EU). *As consumption taxes account for the largest share of low-income groups in society,* as most of their income is spent on current consumption (for basic needs), the poorer people pay more than those on higher incomes do. To ease the regression of consumption taxes, some countries are reducing them for basic household consumption basket goods. Of course, the higher–income population also benefits from such state subsidies, and the problem of inequality is not being addressed again. In this case, measures that are more effective are cash benefits for a certain target group or vouchers for the purchase of basic goods for a certain group of society.

The common employment report of the EC (2015) states that excessive taxation of labor exists in many EU countries. In 2011, the average tax wedge in the EU was 39.6%. (29.5% in the United States), which discourages low-skilled workers and has a negative impact on employment. In 2012, the size of the tax wedge ranged from 20% up to more than 45%. From 2012 to 2013, personal income tax for high-income earners has been increased in 11 member states. The changes in the overall tax wedge were mainly driven by an increase in personal income tax. Assessing personal income tax and social security contributions in general, the burden on employees has increased.

The case of Lithuania. In Lithuania, labor is taxed more than capital and property: the basic income tax rate is only 15%, corporate income tax is 15%, the value-added tax is 21%, and *Sodra's* tax is 30.89%.

The 15% personal income tax in Lithuania is one of the lowest in the EU. Only Bulgaria (10%) applies a lower rate of this tax. By comparison, 21% tariff is applied in Estonia and 25% in Latvia. The highest personal income taxes are in Sweden (56.4%), Belgium (53.7%), and the Netherlands (52%). Since 2000, the amount of personal income tax in Lithuania has decreased by 18% points (from 33%).

The corporate income tax rate in Lithuania (15%) is also among the lowest in Europe. Income tax has been declining since 1995 – from 29% up to 15% (applicable now). EU Member States' corporate tax rates vary several times, from 10% applicable in Cyprus and Bulgaria up to 35% in Malta. Lower rates are in Bulgaria and Cyprus (10% each) and Ireland (12.5%). In Latvia, as in Lithuania, a 15% income tax rate is applied; in Estonia – 21%. The highest corporate tax rates are in Malta (35%), France (34.4%), and Belgium (34%). Capital and profit taxation in Lithuania is lower than the EU average. The average EU corporate tax rate is 23.2%.

Comparing Lithuanian and US capital taxes, the United States pays about 40% of the highest income taxes, while Lithuania – just 15%. However, according to experts, due to various reservations, this number drops to 7% in Lithuania, which shows that corporate and capital taxes in the country are low.

The size of the value added tax (VAT) rate applied in Lithuania (21%) is close to the EU average (20.7%). The lowest 15% VAT rate is in Cyprus and Luxembourg, and the highest – in Denmark, Sweden, and Hungary (25%). Compared to the Baltic countries, Latvia applies 22%, Estonia – 20%. The actual consumption tax burden, calculated by dividing all consumption taxes by household consumption expenditure, was 16.5% – one of the five lowest rates in Europe.

According to Eurostat data, Lithuania collects the lowest amount of taxes on GDP in the EU (according to 2015 data, 29.4% of GDP, while the euro area average is 41.4%). This shows that the tax system is inefficient and the state is unable to insure people, pay adequate social benefits and pensions, and causes other problems, for instance, the shadow labor market.

Lithuania is one of the few EU Member States with a proportional income tax system. The proportional income tax system operates in seven of the 27 EU countries, (Bulgaria, the Czech Republic, Slovakia, Estonia, Latvia, Lithuania, and Romania). In all the other 20 countries – a progressive income tax system has been introduced. It should be noted that all progressive tax systems are very different; for instance, from two income groups in Ireland, Denmark, Poland, and Hungary to 17 groups in Luxembourg. Different progression curves (in some countries, the progression is higher, in others – smaller) and the highest tax rates are also noted (from 30% in Cyprus, 32% in Poland and Hungary, up to 55.9% in Denmark, and 56.6% in Sweden).

Comparing income inequality between EU countries applying a proportional and progressive tax system, for instance, the Gini index of the 7 "proportional" EU countries with the 20 "progressive" countries shows that the average of the Gini index in countries with a progressive tax system is significantly better than the average of the remaining seven EU countries. It can therefore be concluded that a progressive tax system helps to reduce economic inequality.

The benefits of progressive taxation are primarily based on lowering taxes on the lowest earners and strengthening the middle class. However, progressive taxation needs to be very well thought out. Taxing those on higher incomes could only slightly increase the income of the rest, as the share of more earners is too small to collect significantly more taxes and redistribute them to the poor through social policies. It should be borne in mind that once a certain threshold is reached, taxpayers are no longer interested in paying taxes and try to avoid them, so higher taxes lead to shadows and lower government revenues in the long run.

Therefore, in the interests of social justice, the credibility of the tax system and, at the same time, public satisfaction, the tax system must be reviewed comprehensively, including the taxation of personal income, real estate, and land, assessing the socio-economic efficiency of taxable objects and tax rates. Otherwise, incorrectly selected taxable objects and inaccurately chosen tax rates can cause economic damage or benefits that are disproportionate to the additional administrative costs of collecting the tax in question.

When adjusting the tax system, it is necessary to pay attention to the extent to which it will affect business conditions and investments, because it is a business that creates jobs and provides income for the population. Therefore, tax policy needs to be predictable, its impact calculated and assessed in advance. According to R. Kuodis, a good tax system must comply with five principles: horizontal and vertical justice, maximum efficiency, the cheapness of administration, transparency, and flexibility.

According to the authors, the implementation of social justice in Lithuania must take place by adjusting the tax policy in the following directions: *by introducing a progressive tax system and by consistently implementing the policy of increasing property, capital, and environmental taxes.*

Revenue redistribution rates. In creating a welfare state, it is necessary to understand how the income of the social strata should be regulated to encourage the population to work creatively and intensively, realizing their abilities and opportunities. However, it is also important to take into account very different factors that limit those opportunities.

Many of the criteria for assessing economic inequality that has already been examined are merely measures of that inequality, which do not answer the question of how that inequality should be reduced. There is only one criterion that indicates the direction of reducing inequality, and that is the Robin Hood Index. This index roughly shows the share of total income that households with above-average income should transfer to households with below-average income to distribute income evenly (Rudzkienė, 2005). The size of the index is the maximum distance between the line and the Lorentz curve and shows the maximum share of total income that should be redistributed to implement the principle of equality. Putting this index into practice is quite difficult, but it can be a guide for politicians and economists.

The calculations of the Robin Hood Index values performed by the researchers of the MRU Quality of Life Laboratory were helped to determine what proportion (25.96% of total revenue) needs to be redistributed so that incomes are evenly distributed across all sections of the population. The Lorentz curve also needs to set specific levels of income redistribution to shape specific programs to reduce income inequality. Thus, in the outcome of these calculations, as a means to implement the principle of equality and reduce the differentiation of the income of the population, the authors propose tax measures targeting certain groups of society, for instance, the extent to which it is appropriate to differentiate tax rates for those receiving income.

4.2 Macroeconomic policies helping to promote full employment

One of the goals of the EU's Europe 2020 strategy is full employment and social cohesion, which have never been achieved. On the one hand, the prospects for employment growth depend on the EU's ability to promote growth and efficiency through pro-macroeconomic policies (reducing inflation, interest rates, and government deficits). On the other hand, they depend on appropriate microeconomic structural policies (investment, business, and industry), which would create favorable conditions for reducing inequality – increasing the income of the population, reducing poverty and exclusion, and raising the quality of life of the population. Policies to reduce inequality should not only help the economy to recover in the short term but also ensure the necessary long-term social investment to increase people's incomes.

A successful solution to economic inequality cannot be limited to social policy measures. Reducing inequality requires, in particular, reducing unemployment and increasing employment.

This is clearly understood and emphasized in the global strategy papers, including those of the EU. The EC emphasizes the need for a friendly monetary policy for employment policy, and the negative effects of macroeconomic (tight monetary and fiscal) policies on the labor market during the crisis and the post-crisis period are very clear in the International Labor Organisation (ILO) documents. The impact of austerity policies and their negative consequences for the labor market and employment policies are underlined by the ILO's 2013 report Global Employment Trends: "Lack of coordination between monetary and fiscal policies has increased labor market uncertainty" – which means that the priority of monetary and fiscal policy in the neoliberal model of monetary economic development reduces employment and relentlessly increases unemployment. Austerity measures, according to ILO Director-General Ryder, "influenced employment, weakened the influence of the social partners, and harmed a social dialogue. In the context of growing income inequality and the number of poor, these phenomena pose a threat".

The impact of tight fiscal discipline on income and employment. Faced with a state budget deficit, in 2010 many countries embarked on austerity measures, the main ones being expenditure cuts rather than revenue increases. It should be noted that fiscal austerity measures are leading to a sharp decline in investment and social programs.

Austerity policies and reforms aimed at labor market liberalization are failing to stimulate private investment amid weak aggregate demand and financial sector instability. Efforts are being made to counter this trend with even tighter austerity policies, which are leading to a vicious circle that continues to have a negative impact on employment and job creation.

The decline in job opportunities for the unemployed is particularly pronounced in countries with tight fiscal policies: in Spain, the chances of finding a job have fallen from 50% up to 30%, and, in Greece from 25% to 15%. In the countries that have pursued softer fiscal policies, the Czech Republic and Estonia, the chances of finding a job have increased.

Decreasing wages reduce aggregate demand and consumption, which in turn reduces employment opportunities and increases inequality. It is equally important that wage developments continue to be consistent with the need not only to correct external imbalances but also to reduce unemployment and inequality.

Solving problems of employment and economic inequality requires employment-friendly macroeconomic policies. The aggregate demand deficit is holding back faster recovery in global labor markets. The ongoing fiscal consolidation in many advanced economies is hampering faster output growth, in addition to weak private sector consumption. The ILO's Global Employment Trends 2014 report highlights that macroeconomic policies and rising labor incomes could significantly improve labor market prospects. Modeling results suggest that in high-income countries such as the G20, adequate income distribution could reduce unemployment by 1.8 percentage points by 2020, equivalent to 6.1 million new job places. These achievements justify the objectives of a softer fiscal policy.

The position of the authors – Lithuania must abandon its dogmatic fiscal policy – to reduce the budget deficit by any means, as usual, by reducing expenditures on wages and social policy. The middle class needs to recoup its contribution to GDP: in a strategic perspective, without rapid growth in wages and incomes, it will not be possible to increase employment and stop inequality.

In EU countries, it is appropriate to pursue *macroeconomic policies aimed at increasing full employment*, ensuring decent jobs, expanding employment opportunities, promoting the adoption of the necessary measures in the labor market, business addressing the effects of the crisis, and, at the same time, ensuring the stability of public finances and the tax base.

Firstly, reducing economic inequality requires finding the right balance between economic growth, job creation, public income, and public budget policies. This means drawing up realistic public spending plans aimed at creating jobs and, at the same time, pursuing fiscal targets. Such measures may include the introduction of a progressive tax system, incentives for low-income households, an increase in the collection of tax revenue, and an increase in the tax base. Fiscal consolidation policies must go hand in hand with active labor market policies (ALMPs).

Second, companies are forced to operate in conditions of weak demand and uncertain prospects. For job-creating growth to be real, it is necessary to make efforts to promote stable growth for companies and to strengthen their job creation potential.

There is an urgent need to find ways to boost lending to small- and medium-sized enterprises (SMEs), which are the largest source of employment and the biggest source of employment. Governments need to look for incentives that encourage larger companies to invest in productive activities and create jobs.

Thirdly, there is a need for policies that promote productivity, technological change, especially green technologies, and professional skills in this area. This is crucial for the efficient transformation of the economy, the transition to the production of organic products in Lithuania, and would ensure economic growth, the contribution of agriculture to GDP, and the creation of jobs.

Such a policy should focus on:

- increasing productivity, competitiveness, and employment;
- promoting diversification into higher value-added industries; and
- the creation of green industries and green jobs.

Stable economic growth, reducing inequality: a combination of macroeconomic, employment, and income policies. Social dialogue between the real economy, the state, and society is crucial for economic growth that reduces inequality. Such a strategy must take into account the existing budgetary and financial imbalances, which can only be overcome through integrated employment and income policies.

Possible measures:

- implement a tax policy aimed at increasing aggregate demand and reducing economic inequality: a progressive tax system, incentives for the poor;
- encourage investment in the real economy: tax measures to encourage private investment (incentives, etc.); innovative public investment to foster job creation (science, R&D); loans to SMEs; and
- prevent wage deflation; increase wages in line with productivity growth.

Although employment in older workers increased during the post-crisis period, it lags far behind overall employment and youth employment rates. Given the aging trend in Central European countries, the labor market situation of older people will deteriorate rapidly. Although the unemployment situation of older people in the EU seems to be better than in other age groups, this does not indicate difficulties in finding a secondary job; their unemployment is longer due to various reasons: less intensive job search, worse employer attitude toward this age group, workers' health constraints, and lack of new skills. The main barrier to secondary employment for older workers is the lack of decent jobs. The chances of finding a job in a secondary job are even lower for the unskilled, the long-term unemployed, women, and the disabled. These trends pose significant challenges for the integration of older groups into the labor market.

Taking into account the recommendations of the ILO and the EC, measures for the retention of older workers in the labor market should combine the following groups of measures that we have singled out and applied to Lithuania:

First, labor market measures to promote active aging:

- The role of ALMPs is to reduce short-term employment schemes or social security contributions for unskilled workers in exchange for maintaining jobs;
- state subsidies, labor exchange subsidies for training, and professional development;
- wage subsidies;
- the transfer of surplus (redundant) workers to new jobs;
- vocational training;
- labor market entry programs;
- public works; and
- self-employment programs.

Second, social protection measures for active aging:

- learning, vocational training, and the development of professional skills (continuous retraining, lifelong learning);

- healthy and safe working conditions are important for prolonging working life;
- reconciliation of working conditions and needs (reduction of workload, reduction of working hours, and part-time employment);
- free access to employment services and ALMPs (secondary employment, individual action plans for older workers, job subsidies, and public works); and
- incentives to hire.

Third, policies to promote economically active aging:

- extending the coverage of ALMPs and the availability of all economic activity programs for the unemployed, while expanding their employability, stimulating their demand in the labor market; and
- the development of economically active aging strategies that promote professional development and retraining, ensure adaptation to new working conditions, the changing needs of older workers, guarantee wide access to individual employment assistance, and promote their employment in existing jobs.

Measures for the retention of older workers in the Lithuanian labor market are recommended taking into account the experience of EU member states:

- *develop a lifelong learning system as a guarantor of survival in the labor market;*
- support initiatives for the development of lifelong learning services based on institutional partnerships; and
- introduce new funding models for non-formal adult education.

Encourage older workers to stay longer in the labor market.

- create favorable conditions for working at retirement age;
- promote the learning of older workers through flexible forms of adult learning;
- to provide the elderly unemployed with the opportunity to update and retrain their skills, to apply targeted incentives for the employment of older workers; and
- create incentives for employers to recruit older workers.

Measures to limit early exit from the labor market:

- limiting of early retirement, tightening retirement conditions, reducing the financial attractiveness of early retirement, and focusing more on measures to promote activity;
- encouraging to work beyond retirement age by setting higher pension benefits;

- increase the retirement age for older workers, gradually increase the retirement age to 70; and
- linking retirement age to life expectancy.

Women are one of the most vulnerable social groups in the EU labor market. The EU and Global Strategy Papers on Gender Equality emphasize the need to address discrimination and segregation in the labor market. Women are employed in lower-paid and less prestigious farm activities and earn on average less than men do. Discrimination against women already occurs when looking for work: the ways in which women and men find work differ, the time they look for work, and the reasons for losing it. Women make up a significant proportion of the long-term unemployed. Unemployment lasts longer for women than for men, and they are more often pushed into the household: women are forced to change their unemployment status to house-wives without losing hope of finding a job. A large proportion of women who are pushed out of the market lose their jobs permanently. These tendencies are also characteristic of Lithuania.

In order to increase women's employment opportunities in Lithuania, it is expedient to implement the following directions and measures of action:

- reconciling work and family life;
- improving the childcare system; and
- improving the business environment for women.

4.3 Social policy measures

Universal basic income model. As one solution to the problem of low income, several European countries have opted for the Universal Basic Income (UBI). The need for basic security has been highlighted by research-ers Baldwin (Switzerland), Sandel (USA), and Standing (UK) for several decades.

In Europe and Canada, this social policy measure is still being experimented with only in individual cities, for instance, in Utrecht, the Netherlands, in Lausanne, Switzerland, even thought that the population did not support the idea during a referendum in Switzerland (presumably due to the relatively high expected universal base salary of EUR 2,000). In 2017, Finland also launched a two-year pilot project, offering to pay EUR 805 per month to each resident of the country (during the experiment – only to selected par-ticipants in the project).

The idea of this model is to pay a fixed amount of money to the basic needs of the population every month, regardless of their income and social status – a resident does not have to meet any criteria in order to receive a universal basic salary, which is typical of ordinary social policy (a resident must be covered by social insurance, have a certain length of service to receive an appropriate old-age pension, etc.).

Both in Finland and in other European countries, the aim is to find out, on the one hand, whether this measure would help to reduce social exclusion and poverty; on the one hand, it would compensate for the administrative burden of social benefits; and, on the other hand, it would encourage the unemployed to return to the labor market by taking up self-employment.

The introduction of these benefits would reduce the administrative burden, as 42 types of different social benefits are currently applied in Lithuania. However, in assessing the effectiveness of this measure, its size needs to be very balanced; as otherwise, the payment of a universal basic salary to those who do not participate in the labor market must maintain or increase the employer's initiative or motivation to work.

Those who implement this model claim that the payment of benefits that provide basic needs would just liberate those who want to work according to their qualifications and hobbies, as they would not then be forced to take up any subsistence work.

The advantages of this model are even simpler: administration (all residents of the country receive income, there is no need to justify the fulfillment of certain criteria), everyone pays taxes to the budget, compulsory insurance taxes, purchasing power, and consumption (of the most deprived class). These advantages increase public confidence, a declining shadow in the labor market, as the population is no longer interested in hiding income due to both tax evasion and imminent loss of benefits above a certain amount (in Lithuania, those who now want to receive social benefits must prove that they have insufficient income; therefore, if such a benefit were to be paid after starting work, the salaries paid in "envelopes" would be reduced).

Nevertheless, in the case of Lithuania, we need to assess the state's financial capacity to pay a basic universal salary, the labor market's ability to pay a competitive wage to make the unemployed interested in employment, and the behavior and mentality of the Lithuanian population (Table 4.1).

According to the current level of expenditures on social support and social insurance in the Lithuanian state and municipal budgets, another alternative would be more acceptable for Lithuania – to apply not a full basic income model, when the universal basic salary replaces all social benefits, but a partial basic income model, when all social benefits are waived and social security benefits are retained.

However, in order to ensure that individuals (families) have sufficient resources to meet their minimum needs, the methodology for calculating a person's (family's) minimum needs must be approved and the number of social benefits must be linked to the level of resources needed to meet minimum needs – based on real consumption costs and purchasing power, and reviewed and indexed annually.

According to the data of the Lithuanian Department of Statistics, about 15% of the population of the country received less than 250 EUR in 2016. According to a survey conducted in 2006, 17% got less than 102 EUR. However, 7.9% indicated that they were living on unemployment benefits

Table 4.1 Indicators of the use of pensions, compensations, and social assistance

		Number (1,000 inhabitants)	Share of total population (per cent)	Expenditure on pensions/benefits/ compensations (EUR million)
	Population of Lithuania (2017)	2,849		
1.	**Pensions**			
	Number of pension recipients (2017)	980	34.4	2,733
	Old-age pensions			*1,997*
	Invalidity pensions			*498*
	For state pensions			*130*
	Widows "and orphans" pensions			*108*
2.	**Social security**			
	Number of recipients of social benefits (2015)	114.7	4.0	296
	Number of students receiving free school meals (2014)	90.5	3.2	17
	Number of recipients of housing heating compensation (2016)	129.8	4.6	17.9
3.	**Unemployment benefits**			
	Number of registered unemployed (2017)	148.6	5.2	138
Total (2 + 3):		**483.6**	**17.0**	**468.9**
Total (1 + 2 + 3):		**1,463.6**	**51.4**	**3 201.9**

Source: Data published by Statistics Lithuania, Lithuanian Labor Exchange, Ministry of Social Security and Labor of the Republic of Lithuania (2017).

or disability pensions. And, 29.6% of the interviewed recipients of old-age pensions indicated that they did not receive 250 EUR. Therefore, it can be reasonably stated that almost one-fifth of the Lithuanian population is the most vulnerable social group that cannot meet the minimum needs from income. In order to achieve and guarantee a more equitable distribution of income and wealth, social policy measures must be targeted primarily at this: target group.

Policies of social benefits which encourage the return to work. State support for the working-age population must be assessed in terms of whether it reduces incentives to work. If the amount of social support benefits received by a person (family) (social benefits, social support for students, child benefits, etc.) is slightly lower or similar to income from work, such a support is unlikely to encourage the population to work. Unfortunately, not all residents realize that social support is not intended to ensure the long-term social security of a resident – it helps a person (family) to survive in difficult conditions, but the supported person should not be accustomed to working, taking care of family, and thinking about the future and saving.

Because of the increasing abuse of social benefits and allowances in recent times, we would suggest that the requirements for access to social benefits and allowances be thoroughly reviewed and, if necessary, tightened. Such conditions may, of course, lead to an even higher level of poverty for some time in the group of beneficiaries, but, on the other hand, returning to the labor market is a necessity rather than a free choice for the beneficiary. Therefore, *the authors propose to set the minimum amount of benefit but to apply the procedure of decreasing benefits, when the period of payment of benefits would be limited and encouraged or even forced to return to the labor market.* However, in order to avoid further exacerbation of inequalities and situations where those who are unprepared for the labor market may be left without basic assistance, as many cases as possible need to be analyzed and regulated, giving priority to measures to move from the unemployed to employee status. Understandably, the cost of implementing such a targeted social support model, as opposed to the universal basic pay model, is rising.

Experience in other countries has shown that *effective benefit policies need to be combined with active recruitment and retraining schemes* (for instance, a multi-professional network based on a one-stop-shop principle in Sweden with social support and employment agencies; Denmark has adopted a law on municipal activation, which requires municipalities to implement the "active recipient of social assistance" model, providing various in-service training and other services to beneficiaries).

In many cases, loss of income in the event of unemployment means that a person is choosing illegal work to survive. And, this becomes completely justified because, e.g., with a lower professional qualification, it is usually likely that only a minimum wage will be offered. In this case, *we would suggest improving the social support system for active returnees rather than passive recipients of social benefits*; for instance, providing various benefits, covering the minimum costs of maintaining housing, and compensating – the cost of training or retraining, thus increasing the incentives of the unemployed to look for work. In this way, those who are not motivated to look for work because they receive income from the shadow economy should "separate".

According to a study by researchers of the MRU Quality of Life Laboratory (2006 survey), 43.8% of respondents justify not paying taxes to the state and a similar share of the population believes that without additional money it would be impossible to live normally. The 29.9% of respondents also admitted that they had paid in an "envelope" for the work or services, and about a fifth of the population admitted receiving some illegal income. Thus, the size of the informal economy in terms of wages alone is high and distorts official income statistics.

Attention should be paid to the tax burden faced by a person transitioning from social assistance or unemployment status to the labor market. In Lithuania, this tax burden is quite high, so if a person who wants to get a job cannot expect at least 1.5 times more income after taxes, then during the period of "inactivity", his motivation to work will be low. Thus, such taxation of labor income, which practically equates the amounts of social benefits and wages, does not create incentives to work. Therefore, *the authors propose to review and reduce the income tax rates for those entering and/or returning to the labor market in order to encourage the population to move from and remain inactive from social benefits and allowances to the labor market.*

Based on the experience of other countries (Austria, the Czech Republic, Germany, Luxembourg, France, Poland, etc.), we propose to combine pensions and unemployment benefits with earned income. This could be done, for example, by setting an income ceiling on pensions or unemployment benefits, or by setting maximum working hours, without restricting access to the active aging population and young people entering the labor market.

Equal opportunities for access to public services. One of the indirect ways to ensure economic inequality is considered by the authors to be the provision of equal public services to the entire population, which is implemented through the reduction of corruption and nepotism. An example of the successful implementation of such measures is the Scandinavian countries, followed by Estonia. These countries have invested in measures to increase transparency, fight corruption, and strengthen the quality of governance, thus guaranteeing equal access to quality education, health care, and other public services for their people., If the state is corrupt and opaque, then it depends a lot on what you know, who your parents are, in what family you grew up, and whether you are a man or a woman, then the same economy has a negative effect on it. With equal access to public services, the social capital accumulated by the population depends only on its own initiative and personal qualities. Thus, equalization of access to public services – in the fields of education, health care, and infrastructure development – is also one of the important components in reducing economic inequality, increasing the population's trust in state institutions, and satisfaction in the services provided by the state.

4.4 Reducing social vulnerability and building resilience

In recent decades, special attention has been paid to research into vulnerability and resilience. It has been found that the more resilient a population is, the higher their well-being. Unfortunately, few such studies have been conducted in Lithuania, not to mention the links between vulnerability and resilience and economic inequality.

Researchers at the MRU Quality of Life Laboratory 2016 studies show that over 80% of the Lithuanian population face the feeling of insecurity and instability because of the threat of unemployment, falling living standards, economic inequality (lack of social justice), the decline in moral-ethical values, and inefficient state policy.

UNDP 2014 The Human Development Report emphasizes that real progress in human development depends not only on one's choice and ability to get an education, to be healthy and to achieve a certain standard of living, and to feel safe, but also to ensure that these outcomes are stable and sufficient for sustainable human development (Human Development Report 2014). To describe these dependencies, the concepts of human *vulnerability and resilience to life difficulties were introduced in this report.*

The concept of social vulnerability has come into use primarily in the context of natural hazards and disasters (floods, earthquakes); so, it is common for this concept to be frequently used in the risk management literature. Recently, however, social vulnerability has also been understood as an existing condition that describes the social conditions affecting a person (society) and the ability to recover from life difficulties.

The authors perceive vulnerability in a broader sense – as the potential risk of failure of certain social groups, communities, stressful situations, economic difficulties, natural disasters, climate change, and military conflicts, collectively referred to as socio-economic vulnerability.

Vulnerable social groups include children, young people, the elderly, women, and the disabled. Other social groups may be vulnerable, such as the poor, those in the informal sector, those experiencing social exclusion and at risk of being vulnerable, and migrants.

At different stages of their life cycle, people face all levels of insecurity and different forms of vulnerability. Children, young people, and the elderly are vulnerable from the outset, and the question is, what investments and measures can reduce vulnerability at the most sensitive stages of the life cycle?

Resilience to life difficulties is generally considered an object of research in psychology. Resilience, hardiness (AGS), shows a person's ability to cope with a stressful situation while maintaining internal balance and continuing to be successful. The problem of personality abilities raised by psychologists to cope with stress, illness, increase productivity, and improve the quality of life at work has expanded to the whole theory of resilience to life difficulties, conceptual models that are widely applied in the world.

Resilience to life difficulties in the broadest sense means rapid and effective recovery from a highly stressful situation, constructive behavior in significant adverse circumstances, and "positive psychological ability to recover, recovery after a difficult situation, insecurity, conflict, failure, or increased responsibility".

Resistance is relatively variable and can be strengthened. Resilience was thought to be a rare trait a few decades ago, but it is now recognized as a psychological ability that all individuals possess and can be developed. In other words, every person has the potential for resilience, but the real opportunities for people to use this resource during negative stress and disaster.

Interestingly, there is evidence that the power of resilience, once activated, not only allows an individual to recover from a particular event that has affected him but also to function even more successfully than before, such as a negative event at the workplace level

Vulnerability and resilience issues have been analyzed, evaluation methodologies have been proposed, and comparative analyses based on various composite indicators have been performed by the following foreign researchers M. Gall, J. Birkmann, I. Schauser, E. Tate, B. Beccari, B. Khazai, C. Easter, C Pfefferbaum, and L. Rose. So far, empirical research projects in this area are quite rare, conducted on a very narrow, small whole (athletes, managers, teachers, military, and students). Thus, the phenomenon of social vulnerability and resilience to life difficulties remains little studied to date.

The interaction between socio-economic vulnerability and resilience is characterized by the following groups of factors: (1) vulnerable social groups (who are vulnerable); (2) difficulties, threats, and threats (how society or its social groups are vulnerable); and (3) reducing the risk or consequences of threats through promoting resilience to the challenges of life – educating and educating the public, building a community of solidarity, developing skills for independent living, etc. (see Figure 4.2).

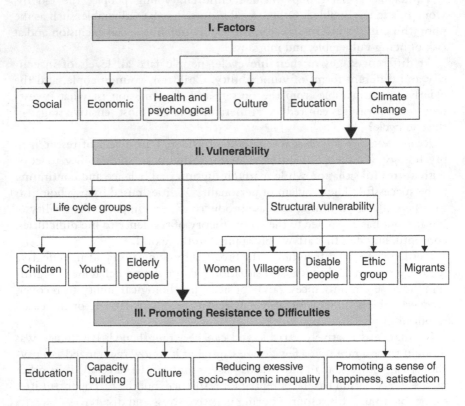

Figure 4.2 A systematic concept of vulnerability and resilience to life difficulties.

Source: Compiled by the authors.

The following factors are very important in building human potential: first, human potential is affected by *investment in science and education* at all stages of the human life cycle. The earlier they are, the greater the human perspective. Conversely, if they are untimely and shortlived, there is a high probability that man will not be able to realize his human potential, to realize himself.

According to Piketty, one of the most important levers for convergence, the reduction, and contraction of inequality, is *the dissemination of knowledge and investment in education and skills development*. The impact of the economic law of supply and demand is less than the dissemination of knowledge and capacity building and is often unclear and contradictory in its consequences. Dissemination of knowledge and capacity building is key to increasing productivity and reducing inequalities both within and between countries. During the research, the authors sought to determine how Lithuanians value the opportunity to increase their income by developing skills in individual areas. The results of the survey showed that the majority of respondents are convinced that their income would not change at all if they increased their skills in any of the listed areas, but the sociological survey revealed that the Lithuanian population would also increase. It should be noted that people with higher education and younger people living in big cities are more likely to believe that developing their skills could contribute to the growth of their personal income.

Second, culture and the value system can affect – increase or decrease the resilience of a person and a country to the difficulties of life. In answering the question of why one country is lagging and the other is progressive, and whether a lagging country can become a leader, more and more scientists in the world (A. Auzan, D. North, and B. Weingast) state that in modern economic theory, there is a "path-dependent effect": a country gets into a named track, tries to break out of it, but all the time slips back into them. From the scholars' point of view, what keeps the country on a lumpy path is not related to economic growth, not to the effectiveness of the economic policy, but to values and norms of behavior: what people consider evil and good, what is acceptable and what is unacceptable. So, the point lies in a culture that can change depending on education, training, and long-term work with people.

Third, public policy and society can empower people to overcome obstacles and threats, but one of the factors that "contributes most" to increasing vulnerability is excessive inequality, especially when the poor are unable to cope with the difficulties of life.

The ability of the society to overcome difficulties is significantly increased by the economic policy pursued by the state, the attitude of state institutions to the macroeconomic factors determining the well-being of the population – not only employment, tax, and income distribution, social policy but also cultural, educational policy, and psychological factors affecting the socio-economic vulnerability of society through the economic policy of the state.

Table 4.2 Intensity of social vulnerability factors of Lithuanian population (%)

No.	Vulnerability factors	Absolutely no fear	Don't feel fear	Yes and no, in part	Feels fear	Feels great fear	Feels fear in total
1	*2*	*3*	*4*	*5*	*6*	*7*	*6 + 7*
1.	Loved ones disease, fear of loss	5.8	10.6	23.0	38.0	22.5	60.5
2.	Fear of becoming disabled	9.8	14.4	20.2	30.2	25.2	55.4
3.	Fear of deteriorating health, illness	9.2	11.1	27.7	33.6	18.4	52.0
4.	Threat of deterioration of living standards	7.4	9.5	32.0	36.5	14.6	51.1
5.	Threat of loss of revenue	10.7	10.5	30.9	34.5	13.1	47.6
6.	The threat of poverty, deprivation	10.7	15.3	26.8	32.1	15.2	47.3
7.	Fear of old age	16.5	25.4	26.4	20.4	11.1	31.5
8.	Fear of social injustice	16.2	22.8	29.3	21.8	9.4	31.2
9.	Fear of losing your job	42.3	14.1	17.0	16.8	9.4	26.2
10.	Fear of loneliness	26.4	24.4	24.3	17.3	7.6	24.9
11.	Fear of not finding a suitable job	46.0	13.2	18.1	16.4	5.7	22.1
12.	Threat of emigration	49.8	18.9	14.5	10.0	6.3	16.3

Source: (Survey of the Lithuanian population in 2016, $N = 1,001$).

Social vulnerability and resilience of the Lithuanian population to life difficulties. The survey of the Lithuanian population examines the economic, social, and psychological status of vulnerable groups and the possibilities of strengthening resilience to life difficulties through psychological factors, as well as the links between psychological resilience and economic inequality.

A specific expression of vulnerability is people's fear of the declining quality of life: deteriorating health, fear of illness, disability; deterioration of material well-being, and the threat of loss of income; as well as fear of social injustice, old age and loneliness, and so on (see Table 4.2).

The results of the study indicate a high level of vulnerability of the Lithuanian population. Depending on the threat factors – 30–60% of the country's population expresses anxiety caused by various fears and threats.

Health factors are in the first place – as much as 60.5% of the population fears the illness and loss of loved ones, which is a relatively natural phenomenon. Fear of deteriorating health and illness worries 52% of the population. The fear of becoming disabled covered – 55.4% of respondents.

This type of anxiety can be explained by several objective factors. First, a person feels fear of becoming incapacitated because in that case, he loses a real source of income. Social guarantees in Lithuania are one of the smallest in the EU, so a person does not expect to receive any social support. Secondly, the commercialized healthcare system does not provide people with medical care – medicines and health services are expensive and their availability is reduced.

Table 4.3 The biggest threats in the country that increase human vulnerability (%)

No.	Threat characterization	I totally disagree	I do not agree	Yes and no, partly	I agree	Totally agree	Agree in total
1	2	3	4	5	6	7	6 + 7
1.	Unemployment	2.2	2.7	11.2	45.8	38.0	83.8
2.	Ineffective public policy	0.9	1.3	14.1	41.0	41.8	82.8
3.	Poverty	1.2	1.9	15.1	48.5	33.2	81.7
4.	Decline of moral-ethical values, degradation of society	1.0	2.5	15.7	45.6	35.1	80.7
5.	Economic inequality	0.9	2.0	18.6	46.5	31.9	78.4
6.	Climate change	6.5	15.6	38.1	25.6	13.8	39.4

Source: (Survey of the Lithuanian population in 2016, $N = 1,001$).

The second group of threats is the standard of living – the threat of worse material situation, reduced incomes, or poverty, which is felt by 51.1%, 47.6%, and 47.3% of the population, respectively.

The third group of threats includes the fear of social injustice (31.2%). It is economic inequality that is likely to hurt and negatively affect more than a third of the population, making them feel the threat of social injustice, especially since Lithuania is one of the "leaders" in the EU according to this indicator.

However, according to the majority of the Lithuanian population, the biggest threats in the country that increase vulnerability (see Table 4.3) are unemployment (83.8%), inefficient state policy (indicated by 82.3% of respondents), and poverty (81.7%).

The decline of moral-ethical values and the threat of degradation of society are noted by 80.7% of respondents, and the threat of economic inequality – by 78.4% of the population. According to the hierarchy, the population ranks last for the threat of climate change (39.4%).

The results of the study on resilience to life difficulties (see Table 4.4) show that Lithuanians are actively involved in shaping their lives. This, in contrast to the threat study, is an optimistic picture. Over 60% of Lithuanians react calmly and without panic to the current difficulties and try to do what depends on themselves: 63% – tries to solve problems that arise "step by step", calmly; 66.2% – does everything that depends on them and continues to leave events to their own devices; 69.7% – calmly responds to stress and gradually solves problems. Helplessness overcoming life's difficulties accounts for only about 20% of the population: postpones solving problems, only 17.6% do not fight people; 20.7% – do not want to act. About 42.3% of the population sees the emerging difficulties of life as life challenges that need to be overcome without fear, by checking oneself, gaining experience, and developing one's abilities.

It should also be borne in mind that a significant part of the population partly agrees with the provisions proposed in the study and partly does not.

Table 4.4 Resistance to life difficulties (%)

No.	A statement describing resilience to the difficulties of life	I totally disagree	I do not agree	I agree in part, not in part	I agree	Totally agree	Agree in total
1	2	3	4	5	6	7	6 + 7
1.	I react badly to problems, they break me, paralyze the initiative, do not make me want to act	8.2	31.1	40.1	18.4	2.3	20.7
2.	I react very sensitively to problems and difficulties in life, which causes me a lot of tension, but when I get up I try to do something, act	5.3	19.4	39.8	32.0	3.6	35.6
3.	I try to calmly solve the problems step by step	1.4	6.8	28.8	54.9	8.1	63.0
4.	The key is to respond calmly to stress and gradually resolve any issues that arise	1.0	5.8	23.5	58.9	10.8	69.7
5.	I do everything that depends on me, and then—as it will be, it will be so	1.4	3.9	28.5	56.5	9.7	66.2
6.	I postpone the solution of problems for a further period, I do not fight, life will show, it will solve them itself	6.0	34.2	42.3	16.6	1.0	17.6
7.	I see the challenges that arise as life challenges that need to be met without fear, they help test myself to see if I can overcome it	2.8	13.4	41.3	37.0	5.6	42.6

Source: Survey of the Lithuanian population in 2016 ($N = 1,001$).

In this way, it can be hypothetically stated that there are even more resilient and viable people in Lithuania.

The authors sought to identify the main threats and factors leading to its vulnerability, to assess the population's psychological resilience and ability to overcome difficulties according to different groups of subjective socio-economic stratification and income quintiles and to determine how economic inequality is related to personal resilience.

Global research and our research have shown that people with high resilience also have a high level of well-being, are better prepared to meet the demands and challenges of a changing environment, are open to new experiences, are more emotionally stable, and face life difficulties. Resistance is not genetically determined, it is not an immutable, innate human trait, and it means that it can be altered, strengthened and nurtured. Accordingly, the level of subjective well-being can change through change and the development of human resilience. Research suggests that special interventions can statistically significantly alter a person's psychological well-being.

This chapter presents a relatively new approach, which proposes to address *economic inequality not by traditional social policy measures, but by strengthening the resilience of the individual and society as a whole*: building an educated, highly cultural and moral, cohesive society; reducing social tensions; developing a happy person who understands the meaning and fullness of life and participates in the development of socio-economic life.

According to the data of the survey of a representative sample of the Lithuanian population, in order to reduce the vulnerability of the Lithuanian population, it is recommended to strengthen the resilience of the society through special interventions.

1 Based on the research conducted by the authors, the practical experience of foreign countries and taking into account the groups most affected by income and property inequality, public authorities are suggested to focus on the following main measures to reduce socio-economic inequality:

- ensuring sufficient social support to enable households to meet basic needs, but at the same time, e.g., the unemployed and recipients of social benefits would not lose motivation to look for work – Possible solutions: indexation of pensions, unemployment insurance, social benefit benefits, and the introduction of basic (full or partial) universal salary for vulnerable social groups;
- differentiated taxation, progressive taxation, reduction of the tax burden on the lowest paid, and reduction of shadow and corruption;
- increasing employment, giving priority to the low-skilled and those living in the regions, whose unemployment rates are significantly higher than those in the high-skilled and those living in metropolitan areas, and should therefore include education, training, retraining, and the facilitation of worker mobility; and
- strengthening the resilience of vulnerable groups to life difficulties (capacity building and psychological intervention measures, taking into account the most common threats and fears experienced by the population).

In welfare states, there are generally two main ways to reduce socio-economic inequality: reforming the tax system and developing a social security system (social benefits policy), which are most effective in redistributing market income.

In the case of Lithuania, a large income differentiation of the country's population is due to the low and even relatively declining income of the population not participating in the labor market – the unemployed, recipients of old-age pensions, and social benefits. According to the structure of the Lithuanian population, one of the largest parts of the total disposable income is accounted for by old-age pensions (the number of recipients is about one third of the Lithuanian population); so, old-age pensions have a significant impact on the overall income distribution. The number of other recipients of social benefits is also not small (about 17–20%), and their share in disposable income is not large due to their size. People living in the poorest and most deprived material well-being do not feel the positive effects of economic growth and, as a result, the gap between the population in the extreme deciles is only widening.

2　Several European countries have opted for a universal basic salary model as one solution to the problem of low income. The idea of this model is to pay a fixed amount of money to the basic needs of the population every month, regardless of their income and social status – the resident does not have to meet any criteria to receive the universal basic salary that is typical of normal social policy. The application of this model in Lithuania would reduce the administrative burden, increase the purchasing power of the most deprived and the population's trust in the government, and reduce the shadow labor market. However, in assessing the effectiveness of this measure, its size needs to be very balanced, as otherwise, the payment of a universal basic salary to those who do not participate in the labor market must maintain or increase the employer's initiative or motivation to work. We must also assess the state's financial capacity to pay a basic universal salary, the labor market's ability to pay a competitive wage so that the unemployed are interested in finding employment, and the behavior and mentality of the Lithuanian population.

3　As part of a policy to encourage the return to work of social benefits, the authors propose a minimum benefit, but a digressive benefit regime with a limited payout period that encourages or even encourages a return to the labor market. However, in order to avoid further exacerbation of inequalities and situations where those who are unprepared for the labor market may be left without basic assistance, as many cases as possible need to be analyzed and regulated, giving priority to measures to move from the unemployed to employee status. Understandably, the cost of implementing such a targeted social support model, as opposed to the universal basic pay model, is rising.

4　Lithuania is one of the few EU Member States with a proportional income tax system. Comparing income inequality between EU countries with a proportional and progressive tax system – the Gini index of 7 "proportional" EU countries and 20 "progressive" countries – we see that the average of the Gini index in countries with a progressive tax system is significantly better than the average of the remaining seven EU countries. It can therefore be concluded that a progressive tax system helps to reduce socio-economic inequalities.

The benefits of progressive taxation are primarily based on lowering taxes on the lowest earners and strengthening the middle class. However, progressive taxation needs to be very well thought out. Taxing those on higher incomes could only slightly increase the income of the rest, as the share of more earners is too small to collect significantly more taxes and redistribute them to the poor through social policies. It should be borne in mind that once a certain threshold is reached, taxpayers are no longer interested in paying taxes and try to avoid them; so, higher taxes lead to shadows and lower government revenues in the long run.

According to the authors, the implementation of social justice in Lithuania must take place by adjusting the tax policy in the following directions: by introducing a progressive tax system and by consistently implementing the policy of increasing property, capital, and environmental taxes.

5 The authors' calculations of the Robin Hood index values made it possible to determine what proportion of total revenue (about a quarter of the population) needs to be redistributed so that incomes are evenly distributed across all sections of the population, and the Lorentz curve needs to set specific levels of income redistribution to shape specific programs to reduce income inequality. In the outcome of these calculations, as a means to implement the principle of equality and reduce the differentiation of the income of the population, the authors propose tax measures targeting certain groups of society – for those who receive income, it is appropriate to differentiate tax rates, what proportion of total income needs to be redistributed among population groups in order for the income to be distributed equally in all sections of the population. For example, an additional 7.5% should be allocated to the first interval and a reduction of 16.5% to the tenth interval gross income. The authors' calculations show that even up to and including the seventh interval – the income of the population receiving "in the hands" every month up to 901–1,100 EUR should be increased. One of the proposals could be to reform the tax system by introducing progressive taxes between different income groups, based on the percentages of redistribution of total income provided in the study.

6 One of the main directions to solve the problem of inequality in Lithuania is to implement macroeconomic policies focused on increasing employment (as opposed to austerity macroeconomic policies), which would ensure the creation of decent jobs, expand employment opportunities, and promote business while ensuring the stability of public finances. Leverage to implement such a policy:

• the right balance between economic growth, job creation, population growth, and public budget policies (fiscal consolidation policies must go hand in hand with ALMPs);
• job-creating growth based on stable business growth, strengthening their job creation potential, in particular by activating credit for SMEs; secondly, by encouraging large companies to invest in productive activities and create jobs; and
• promoting productivity growth, technological change, especially in the field of green technologies, through the development of professional

skills in this field. Such policies must focus, first and foremost, on increasing productivity, competitiveness, and employment; secondly, the promotion of diversification into higher value-added industries; thirdly, the creation of green industries and green jobs.

7 According to Piketty, one of the most important levers for convergence – the reduction and contraction of inequality, is the dissemination of knowledge and investment in education and skills development. The impact of the economic law of supply and demand is less than the dissemination of knowledge and capacity building and is often unclear and contradictory in its consequences. Dissemination of knowledge and capacity building is key to increasing productivity and reducing inequalities both within and between countries. During the research, the authors sought to determine how Lithuanians value the opportunity to increase their income by developing skills in individual areas. The results of the survey showed that the majority of respondents are convinced that their income would not change at all if they increased their skills in any of the listed areas, but the sociological survey revealed that the Lithuanian population would also increase. It should be noted that people with higher education and younger people living in big cities are more likely to believe that developing their skills could contribute to the growth of their personal income.

8 The authors emphasize the nontraditional approach, according to which the problem of socio-economic vulnerability is addressed not by social policy but by strengthening the resilience of people and society to life's difficulties: firstly, reducing excessive inequality and social tension and secondly. The authors propose to strengthen society's resilience to life's difficulties, to develop the abilities and certain psychological qualities of a person, which ensure the stability and reliability of people's choices now and in the future, enabling them to better cope with and adapt to negative phenomena. It is the increase of choice, competence (knowledge and professionalism), and strengthening of psychological qualities.

CONCLUSIONS

1 In recent decades, the growing economic inequality in the global economy has reached unprecedented proportions, and this is confirmed by world research, reports of international organizations, speeches of prominent world economists, and state leaders. Two main lines of research emerge in this area: the first states that economic inequality is justified as a result of a market economy and must be; second, argues that inequality is a problem of the economic system, especially when it escalates into excessive inequality that negatively affects socio-economic progress. The current level of economic inequality in the world raises serious doubts as to its validity.

2 A differentiated approach is proposed to identify when economic inequalities begin to negatively affect countries' socio-economic progress. The attitude of the authors of this book is that economic inequality can be normal (reasonable) and excessive (unreasonable). A certain degree of inequality is justified, depending on education, qualifications, responsibility, as well as on the incentives created for innovation and entrepreneurship, on development, competition, and investment promotion in the country. However, growing economic inequality becomes a problem when it restricts individuals' educational or professional choices, reduces opportunities for self-realization, when individuals' efforts are directed only to meet the most basic needs. Excessive inequality is not just deep inequality, but one that, from a certain level of achievement, begins to hamper socio-economic progress of countries.

3 The results of the comparative analysis of income differentiation in the EU countries assessed by using three methods: according to the decile income differentiation coefficient, the optimal poverty line, and the Gini coefficient allowed to draw the following conclusions.

- Excessive inequality is one that occurs in an unequal income distribution when the decile income coefficient Kd (the ratio of the income of the richest 10% of the population to the income of the poorest 10%) is equal to is 10. Calculations have shown that normal inequality would

DOI: 10.4324/b22984-6

exist if incomes were more evenly distributed and the incomes of the poor in the lower deciles were increased so that Kd fell to 7.

- In EU countries, normal and excessive inequality are assessed based on the assumption that there is a certain optimal poverty line at which inequality is justified. Based on this criterion of the optimal poverty line, normal inequality is found when this indicator does not exceed 15.6–16.5%. The inequality that exceeds this set optimal poverty line is unjustified and can be treated as excessive inequality.
- Excessive inequality is interpreted as a situation where inequality in a country exceeds the optimal one. In this context, it is assumed that the normal/optimal inequality is best reflected in the Gini coefficient of the Scandinavian countries and Denmark in terms of gross disposable income, which in 2019 ranged from 26.2 to 27.6. If the Gini coefficient exceeds these values, excessive inequality occurs.

The results of the conducted study show that the EU countries are divided into three clusters according to the ratio of normal to excessive inequality: countries with high excessive inequality, countries with medium and low inequality, and countries with normal inequality. Applying all three assessment methods, the countries of the first cluster – the countries of high excessive inequality – fall into the "anti-leader" countries sub-category: Bulgaria, Romania, Lithuania, Latvia, Spain, and Italy. The countries with normal inequality are Denmark, Sweden, Finland, the Netherlands, Belgium, Austria, the Czech Republic, Slovenia, and Slovakia.

4 Economic inequality affects the socio-economic progress of EU countries – economic growth, quality of life, and sustainability. The impact of economic inequality on the socio-economic progress of EU countries – economic growth, quality of life, and sustainability – is non-linear. Economic inequality does not have an immediate effect on economic growth, the reaction of society and the economy is "delayed" and becomes apparent in the long run (especially in the third year of economic growth since the onset of inequality). The impact of economic inequality on economic growth varies across EU countries, depending on living standards.

5 Excessive inequality, when the marginal effect of economic inequality becomes detrimental to economic growth, is more pronounced in the group of richer countries. For less affluent countries, it is necessary for gross income to grow. Even if that income growth takes place at the expense of the pour, which deepens the economic inequality of these countries, as a result it mitigates the negative impact of economic inequality on the economic growth of poorer countries. The study confirms that economic inequality is excessive and has a negative impact on the economies of most rich countries, such as Ireland, Denmark, Estonia, Italy, Spain, the United Kingdom, Luxembourg, France, Finland, Sweden, and Germany.

6 The impact of economic inequality on quality of life, measured by the median income indicator, differs for groups of countries in terms of living standards. According to the survey, excessive inequality, where the marginal effect of economic inequality becomes negative and a further increase in economic inequality is associated with a deteriorating quality of life, is evident in more than half of the analyzed EU countries: Bulgaria, Romania, Hungary, Lithuania, Latvia, Estonia, Poland, Italy, Spain, Portugal, Croatia, Greece, Ireland, United Kingdom, Germany, and Luxembourg.

The impact of economic inequality on quality of life, as measured by healthy life years, becomes excessive (the marginal effect of economic inequality becomes negative) when the breaking point, measured by the Gini coefficient on disposable income is reached, that is, 25.34%. According to this indicator, economic inequality has a negative impact on the quality of life in almost all EU countries, as Gini 1 exceeds the breaking point in almost all analyzed EU countries (except the Czech Republic, Slovenia, and Slovakia).

Measuring the quality of life as an indicator of the emigration rate, the results of the study confirm that economic inequality promotes an increase in the emigration rate. Comparing the impact of economic inequality on emigration in the EU countries having higher living standards and in EU countries having lower living standards, the statistical significance of economic inequality indicators in the richer group justifies the situation that economic growth (increase in living standards) largely do not address emigration problems. Therefore, in order to reduce emigration, it is very important to strengthen the ties that keep people in their homeland.

7 Economic inequality has a negative impact on sustainability. However, this effect is different for less affluent and richer countries. The different direction of the marginal effect in the groups of rich and less rich countries can be explained on the assumption that the higher the economic inequality, the lower the negative impact on sustainability. As high economic inequality essentially means a large number of people living in poverty compared to a small number of people living in the rich, the impact on the planet and its sustainability is much smaller. In countries with high economic inequalities, the negative impact on the sustainability of the planet can be largely influenced by the richest members of society. Meanwhile, countries that do not have a very large income gap – economic inequality is not relatively high, are characterized by resource-intensive use, active industrial development and high public consumption, so the negative impact on sustainability is significant. Rich countries have achieved a high level of economic development, a large middle class with high levels of consumption, so growing economic inequality increases environmental pollution, depletes natural resources that

fail to regenerate, and promotes the growing greenhouse effect. All this has a negative impact on sustainability.

8 The negative effects of excessive inequality on socio-economic progress can be avoided through integrated systemic solutions to reduce the level of economic inequality. Empirical evidence confirms that for the EU-28, in order to avoid negative effects of excessive inequality on their socio-economic progress, economic inequality should not exceed 29–30%, measured by the Gini coefficient on disposable income.

9 The authors based on the results of their own research and other scholars research, practical experience in foreign countries, and taking into account the population groups most affected by poverty and wealth differentiation (non-labor market participants, people receiving social assistance, people of retirement age, and rural and urban areas) – propose that the EU institutions should focus on the following key measures to reduce economic inequality:

- Differentiated taxation, application of a progressive tax system, reduction of the tax burden on the lowest earners, and reduction of shadow economy and corruption;
- Increasing employment opportunities, giving priority to the low-skilled and those living in the rural regions, where unemployment rates are significantly higher than those of the high-skilled and those living in cities; therefore, main measures should include education, training, retraining, and the facilitation of worker mobility;
- Ensuring sufficient support to enable households to meet basic needs, but at the same time, e.g., the unemployed and those receiving social benefits would not lose motivation to look for work and work. Possible solutions: indexation of pensions, unemployment insurance, indexing of social benefits based on inflation rate, and introduction of basic (full or partial) universal salary for vulnerable social groups;
- Strengthening the resilience of socially vulnerable groups to life difficulties (capacity building and psychological intervention measures, taking into account the most common threats and fears experienced by the population).

In welfare states, there are generally two main ways to reduce economic inequality: reforming the tax system and developing the social security system (social benefit policy), which are most effective in redistributing market income. However, according to the experience of other countries and researchers I. Joumard, M. Pisu, and D. Bloch, social benefits reduce income inequality more than taxes. Almost 75% of the reduction in income inequality is due to social benefits and the remaining quarter is linked to taxation.

10 Lithuania is one of the few EU Member States with a proportional income tax system. Comparing income inequality between EU countries

applying a proportional and progressive tax system, we can see that the average of the Gini index in the 20 countries with a progressive tax system is significantly better than the average of the Gini index in remaining seven EU countries having proportional taxation systems. It can therefore be concluded that a progressive tax system helps to reduce economic inequality.

The benefits of progressive taxation are primarily based on lowering taxes on the lowest earners and in this way strengthening the middle class. However, progressive taxation needs to be very well thought out. Taxing those on higher incomes could only slightly increase the income of the rest, as the share of more earners is too small to collect significantly more taxes and redistribute them to the poor through social policies. It should be borne in mind that once a certain threshold is reached, taxpayers are no longer interested in paying taxes and try to avoid them; so, higher taxes lead to the growth of shadow economy and lower state budget revenues in the long run.

According to the authors, the implementation of social justice in Lithuania must take place by adjusting the tax policy in the following directions: by introducing a progressive tax system and by rationally implementing the policy of increasing property, capital, and environmental taxes.

11 One of the main directions in order to solve the problem of inequality in Lithuania is to implement macroeconomic policies focused on increasing employment (as opposed to austerity macroeconomic policies), which would ensure the creation of decent jobs, expand employment opportunities, and promote business, while ensuring the stability of public finances. Leverages for implementing such a policy are:

- The right balance between economic growth, job creation, population growth, and public budget policies. Fiscal consolidation policies must go hand in hand with active labor market policies (ALMPs);
- Job-creating economic growth based on stable business growth by strengthening their job creation potential, in particular by activating small- and medium-sized enterprises (SMEs) lending; secondly, by encouraging large companies to invest in productive activities and create jobs;
- Promoting productivity growth, technological change, especially in the field of green technologies, by developing professional skills in this field. Such a policy must focus, first and foremost, on increasing productivity, competitiveness, and employment; secondly, the promotion of diversification into higher value–added industries; thirdly, the creation of green industries and green jobs.

12 In addition, the authors propose a new approach for addressing socio-economic vulnerability not through social policies but by strengthening human and societal resilience to life's difficulties: firstly,

reducing excessive inequality and social tension and, secondly, developing an active, spiritually strong person fullness. The authors propose to strengthen society's resilience to life's difficulties, to develop abilities and certain human psychological qualities that provide people with stability and reliability of choices now and in the future, enabling them to better cope with and adapt to negative phenomena. It provides for the increase of choices, competence (knowledge and professionalism), and the strengthening of psychological qualities.

LITERATURE

Acemoglu D., Robinson J. A. (2013). Why Nations Fail: The Origins of Power, Prosperity and Poverty. Profile Books Ltd., 529 p., ISBN 978-1-84668-430-2.

Aidukaitė J., Bogdanova N., Guogis A. (2012). Gerovės valstybės kūrimas Lietuvoje: mitas ar realybė? Lietuvos socialinių tyrimų centras, Sociologijos institutas, 410 p., ISBN 978-609-95257-6-1.

Alesina A., Perotti R. (1994). The Political Economy of Growth: A Critical Survey of the Recent Literature. World Bank Economic Review, Vol. 8, No. 3, 351–371 pp.

Alesina A., Perotti R. (1996). Income Distribution, Political Instability and Investment. European Economic Review, Vol. 40, No. 6, 1203–1228 pp.

Alesina A., Rodrik D. (1994). Distributive Politics and Economic Growth. The Quarterly Journal of Economics, Vol. 109, No. 2, 465–490 pp.

Alvaredo F., Atkinson A. B., Piketty T., Saez E. (2013). The Top 1 Percent in International and Historical Perspective. NBER Working Paper, No. 19075, 1–14 pp.

Alvaredo F., Chancel L., Piketty T., Saez E., Zucman G. (2018). World Inequality Report. World Inequality Lab. Available online: https://wid.world/document/world-inequality-report-2018-english/.

Andrei A., Craciun L. (2015). Inequality and Economic Growth: Theoretical and Operational Approach. Theoretical & Applied Economics, Vol. 22, No. 1, 177–186 pp.

Armstrong S. (2018). The New Poverty. Verso, 244 p. ISBN-13: 978-1-78663-465-8.

Atkinson A. B. (1997). Measurement of Trends in Poverty and the Income Distribution. Cambridge Working Papers in Economics 9712, Faculty of Economics, University of Cambridge.

Atkinson A. B. (2009). Factor Shares: The Principal Problem of Political Economy? Oxford Review of Economic Policy, Vol. 25, No. 1, 3–16 pp.

Atkinson A. B., Piketty T., Saez E. (2011). Top Incomes in the Long Run of History. Journal of Economic Literature, Vol. 49, No. 1, 3–71 pp.

Ayres C. E. (1944). The Theory Of Economic Progress. University of North Carolina Press, 317 p.

Barro R. J. (2000). Inequality and Growth in a Panel of Countries. Journal of Economic Growth, Vol. 5, No. 1, 5–32 pp.

Barro R. J. (1999). Inequality, Growth, and Investment. National Bureau of Economic Research. NBER working paper series. 52 p.

Barro R. J. (2002). Quantity and quality of economic growth. Banco Central de Chile, 135–162 pp.

Bylund P. L. (2019). Entrepreneurship and Austrian Economics: Theory, History, and Future. Interdisciplinary Journal of Philosophy Law and Economics, Vol. 7, No. 3, 517–521 pp.

Bourguignon F. (2004). The Poverty-Growth-Inequality Triangle. The World Bank Website. Working Paper No. 125. Available online: http://documents1.worldbank.org/curated/en/449711468762020101/pdf/28102.pdf.

Bourguignon F. (2016). Inequality and Globalization. Foreign Affairs, Vol. 95, No. 1, 11 p.

Bourguignon F. (2018). World Changes in Inequality: An Overview of Facts, Causes, Consequences, and Policies. CESifo Economic Studies, Vol. 64, No. 3, 345–370 pp.

Brancaccio E., Fontana G. (2011). The Global Economic Crisis. New Perspective on the Critique of Economic Theory and Policy. London: Routledge, 1–10 pp.

Braun D. (1997). The Rich Get Richer: The Rise of Income Inequality in the U. S. and the World. Wadsworth Publishing, 538 p. ISBN978-0830414338.

Brynjolfsson E., McAfee A. (2016). The Second Machine Age: Work, Progress, and Prosperity in a Time of Brilliant Technologies. W. W. Norton & Company, 336 p. ISBN 78-0393350647.

Chen B. L. (2003). An Inverted-U Relationship between Inequality and Long-run Growth. Economics Letters, Vol. 78, No. 2, 205–212 pp.

Čiegis R., Dilius A. (2012). Ekonominio Augimo Poveikio Darniam Vystymuisi Vertinimo Sistemos. Management Theory & Studies for Rural Business & Infrastructure Development, Vol. 33, No. 4, 22–33 pp.

Čiegis R., Dilius A. (2018). Pajamų Nelygybės Poveikio Ekonomikos Augimui Vertinimas Europos Sąjungos Šalių Grupėse. Regional Formation & Development Studies, Vol. 26, 44–55 pp.

Čiegis R., Dilius A., Štreimikienė D. (2020). Pajamų nelygybės poveikio ekonomikos augimui ir darniam vystymuisi vertinimas Europos Sąjungos Šalyse. Vilniaus Universiteto leidykla, 467 p., ISBN 978-609-07-0374-8.

Čiegis R., Pečkaitienė J. (2013). Darnaus Vystymosi Poveikis Gyvenimo Kokybei. Management of Organizations: Systematic Research, No. 68, 7–26 pp.

Clark C. (1940). The Conditions of Economic Progress. London: Macmillan and Co., Ltd., 504 p.

Coccia M. (2018). Economic Inequality Can Generate Unhappiness That Leads to Violent Crime in Society. International Journal of Happiness and Development, Vol. 4, No. 1, 1–24 pp.

COM (2010). Europe 2020. A strategy for smart, sustainable and inclusive growth. Communication from the Commission. Available online: https://ec.europa.eu/research/innovation-union/pdf/innovation-union communication_en.pdf.

Corak M. (2013). Income Inequality, Equality of Opportunity, and Intergenerational Mobility. Journal of Economic Perspectives, 27 (3), 79–102 pp.

Cowell F. A. (2011) Measuring Inequality. Oxford: OUP Oxford (LSE Perspectives in Economic Analysis).

Dabla-Noris, E., Kochhar, K., Suphaphiphat, N., Ricka, F., Tsounta E. (2015). Causes and Consequences of Income Inequality: A Global Perspective. International Monetary Fund, Vol. SDN/15/13.

Daugirdas V., Baranauskienė V., Burneika D., Kriaučiūnas E., Mačiulytė J., Pocius A., Pociūtė–Sereikienė G., Ribokas G. (2019). Netolygaus regioninio vystymosi problema Lietuvoje: socio-ekonominiai gerovės aspektai/Kolektyvinė monografija. Lietuvos socialinio tyrimų centras, 140 p., ISBN 978-9955-531-66-1.

Davies J. B. at al. (2000). "The Distribution of Wealth," in Atkinson A. B., Burguignon F. (eds.) Handbook of Income Distribution, North-Holland Elsevier, 605–75 pp.

Davies J. B. (2013). "Wealth and Economic Inequality" in Salverda, W., Nolan B., Smeeding T. M. (eds.) The Oxford Handbook of Economic Inequality. Oxford University Press.

Davis H. S. (1947). The Industrial Study of Economic Progress. University of Pennsylvania Press Anniversary Collection, 200 p. ISBN 978-1512811162.

Deaton A. (2010). Understanding the Mechanisms of Economic Development. Journal of Economic Perspectives. Vol. 24, No. 3, 3–16 pp.

Deaton A. (2015). The Great Escape: Health, Wealth, and the Origins of Inequality. Princeton University Press, 360 p., ISBN 978-0-691-16562-2.

Deaton A., Stone A. A. (2013). Economic Analysis of Subjective Well-Being. Two Happiness Puzzles. American Economic Review, Vol. 103, No. 3, 591–597 pp.

Deaton A. (2003). Health, Inequality, and Economic Development. Journal of Economic Literature, Vol. XLI, 113–158 pp.

Dominicis, L., Florax, R. J. G. M., Groot, H. L. F. (2008). A Meta-Analysis on the Relationship between Income Inequality and Economic Growth. Scottish Journal of Political Economy, Vol. 55, No. 5, 654–682 pp.

Dudin M. N., Ljasnikov N. V., Kuznecov A. V., Fedorova I. J. (2013). Innovative Transformation and Transformational Potential of Socio-Economic Systems. Middle-East Journal of Scientific Research, Vol. 17, No. 10, 1434–1437 pp. ISSN 1990-9233.

European Commission. Europe 2020. A strategy for smart, sustainable and inclusive growth. Communication from the Commission. Available online: https://ec.europa. eu/research/innovation-union/pdf/innovation-union-communication_en.pdf.

Eurostat (2020). Gini coefficient of equivalised disposable income. Access through internet: https://ec.europa.eu/eurostat/databrowser/view/tessi190/default/table?lang=en [viewed 2020-05-20]

Eurostat Database (2020). The statistical office of the European Union. Access through internet: https://ec.europa.eu/eurostat/data/database

Fagan H. B. (1935). American Economic Progress. J. B. Lippincott Company, 589 p.

Friedrich R. J. (1982). In Defense of Multiplicative Terms in Multiple Regression Equations. American Journal of Political Science, Vol. 26, No. 4, 797–833 pp.

Galbraith J. K. (2008). Inequality, Unemployment and Growth: New Measures for Old Controversies. Journal of Economic Inequality, Vol. 7, No. 2, 189–206 pp.

Galbraith J. K. (2016). Causes of Changing Inequality in the World. Intereconomics, Vol. 51, 55–60 pp. Available online: file:///H:/b00257/Downloads/causes-of-changing-inequality-in-the-world%20(3).pdf.

Galbraith J. K. (2018). Global Oligarchy. Nation, Vol. 307, No. 2, 17–19 pp.

Ginevičius R., Tvaronavičienė M. (2003). Globalization Processes in Baltic Countries: analysis of Trends in Lithuania, Latvia and Estonia. Journal of Business Economics and Management. North-German Academy of Informatology (Stralsund) e.V. Vol. IV, No. 1.

Grifell-Tatje E., Lovell C. A. K., Turon P. (2018). The Business Foundations of Social Economic Progress. Business Research Quarterly, Vol. 21, No. 4, 278–292 pp.

Gruževskis B., Orlova U.L. (2012). Sąvokos "gyvenimo kokybė" raidos tendencijos. Socialinis darbas, Vol. 11, No. 1, Vilnius: Mykolas Romeris University.

Gruževskis B., Zabarauskaitė R., Stankūnienė V., Sipavičienė A., Jasilionis D., Gaidys V., Lazutka R., Kazakevičiūtė J., Maslauskaitė A. (2012). Lietuvos socialinė raida ekonomikos nuosmukio sąlygomis. Lietuvos socialinė raida. Nr. 1. Vilnius. Spaudmeta. 121 p.

Halter D., Oechslin M., Zweimüller J. (2013). Inequality and Growth: The Neglected Time Dimension. Journal of Economic Growth, Vol. 19, No. 1, 81–104 pp.

Hansen A. H. (1939). Economic Progress and Declining Population Growth. Kessinger Publishing, LLC (2010), 24 p. ISBN 978-1162557823.

Hoeller P., Jourmand I., Pisu M., Bloch D. (2012). Less Income Inequality and More Growth – Are They Compatible? Part 1. Mapping Income Inequality Across the OECD. OECD Economics Department Working Papers, No. 924, OECD Publishing, Paris.

Human Development Report 2014 (2014). Sustaining Human Progress: Reducing Vulnerabilities and Building Resilience. New York: UNDP.

Jackson T. (2012). Gerovė be augimo: ekonomika ribotų išteklių planetai. Vilnius: Petro ofsetas, Grunto valymo technologijos, 271 p., ISBN 978-609-420-261-2.

Jenkins S. P., Micklewright J. (2007). Inequality and Poverty Re-Examined. Oxford University Press, 304 p. ISBN 9780199218110.

Jenkins S. P., Micklewright J. (2007). New Directions in the Analysis of Inequality and Poverty. IZA Discussion Paper No. 2814, Available at SSRN: https://ssrn.com/abstract=995502.

Kaasa A. (2005). Factors of Income Inequality and Their Influence Mechanisms: A Theoretical Overview. Tartu University Press, 48 p., ISBN 9985-4-0460-2.

Krugman P. (1995). Peddling Prosperity: Economic Sense and Nonsense in an Age of Diminished Expectations. W.W. Norton & Company, 303 p., ISBN 9780393312928.

Krugman P. (2009). The Conscience of a Liberal. W.W. Norton & Company, 2009, 296 p., ISBN 978-0-393-33313-8.

Krugman P. (2018). The Return of Depression Economics and the Crisis of 2008. W.W. Norton & Company, 2009, 224 p., ISBN 9780393337808.

Krugman P. (2020). Arguing with Zombies: Economics, Politics, and the Fight for a Better Future. W.W. Norton & Company, 464 p., ISBN 978-1324005018.

Kumhof M., Ranciere R., Winant P. (2015). Inequality, Leverage and Crises. American Economic Review, Vol. 105, No. 3, 1217–1245 pp.

Kuznets S. (1955). Economic Growth and Income Inequality. The American Economic Review, Vol. 45, No. 1, 1–28 pp.

Lagarde C., Stiglits J. E., Brynjolfsson E. et al (2016). GDP a poor measure of progress, say Davos economists. The Annual Meeting took place in Davos from 20 to 23 January, under the theme "Mastering the Fourth Industrial Revolution". Available online: https://www.weforum.org/agenda/2016/01/gdp/.

Lawson M., Chan M. K., Rhodes F., Anam Butt A. P., Marriott A., Ehmke E., Jacobs D., Seghers J., Atienza J., Gowland R. (2019). Public good or private wealth? Oxfam International. Available online: https://indepth.oxfam.org.uk/public-good-private-wealth/

Lazutka R. (1997). Socio-economic Environment of Social Policy Development. EMERGO, Journal of Transforming Economies and Societies, Vol. 4, No. 2, 81–97 pp.

Lazutka R. (2001). Socialinė Apsauga. Žmogaus socialinė raida. Vilnius: Homo Liber, 131–150 pp.

Lazutka R. (2003). Gyventojų pajamų nelygybė. Filosofija, sociologija, Vol. 2, 22–29 pp.

Lazutka R. (2007). Gerovės kapitalizmo raidos problemos Lietuvoje. Lietuvos ekonominė padėtis Europoje ir globalioje erdvėje. Vilnius: Ekonomikos tyrimų centras, 61–82 pp.

Lazutka R. (2014). European Minimum Income Network country report Lithuania. Analysis and Road Map for Adequate and Accessible Minimum Income Schemes in EU Member States. European Commission, 37 p.

Lazutka R., Poviliūnas A. (2009). Lithuania: Minimum Income Schemes. A Study of National Policies. On behalf of European Commission DG Employment, Social Affairs and Equal Opportunities.

Lazutka R., Juška A., Navickė J. (2018). Labour and Capital Under a Neoliberal Economic Model: Economic Growth and Demographic Crisis in Lithuania. Europe-Asia Studies, No. 70:9, 1433–1449 pp.

Lefranc A., Pistolesi N., Trannoy A. (2008). Inequality of Opportunities vs. Inequality of Outcomes: Are Western Societies All Alike? Review of Income & Wealth, Vol. 54, No. 4, 513–546 pp.

Li H., Squire L., Zou H. (1998). Explaining International and Intertemporal Variations in Income Inequality. Economic Journal, Royal Economic Society, Vol. 108, No. 446, 26–43 pp.

Li H., Zou H. (1998). Income Inequality is not Harmful for Growth: Theory and Evidence. Review of Development Economics, Vol. 2, No. 3, 318–334 pp.

Lithuanian Department of Statistics (2019). Household income and living conditions. Statistical yearbook of Lithuania (edition 2019). Available online: https://osp.stat. gov.lt/lietuvos-statistikos-metrastis/lsm-2019/gyventojai-ir-socialine-statistika/namu-ukiu-pajamos-ir-gyvenimo-salygos

Lorenzi P. (2016). Inequality and Economic Growth. Society, Vol. 53, No. 5, 474–478 pp.

Luthans, F. (2002). Positive Organizational Behavior: Developing and Managing Psychological Strengths. Academy of Management Executive, Vol. 16, No. 1, 57–72 pp.

Luthans, F., Avey, J. B., Clapp-Smith, R., Li, W. (2008). More Evidence on the Value of Chinese Workers' Psychological Capital: A Potentially Unlimited Competitive Resource? The International Journal of Human Resource Management, Vol. 19, 818–827.

Luthans, F., Avolio, B. J., Norman, S. M., Avey, J. B. (2006). Psychological Capital: Measurement and Relationship with Performance and Satisfaction. Gallup Leadership Institute Working Paper. Lincoln, NE: University of Nebraska.

Luthans, F., Avolio, B. J., Walumbwa, F. O., Li, W. (2005). The Psychological Capital of Chinese Workers: Exploring the Relationship with Performance. Management and Organization Review, Vol. 1, 249–271 pp.

Luthans, F., Youssef, C. M. (2004). Human, Social, and Now Positive Psychological Capital Management. Organizational Dynamics, Vol. 33, 143–160 pp.

Luthans, F., Youssef, C. M., Avolio, B. J. (2007). Psychological Capital: Developing the Human Competitive Edge. Oxford University Press, Oxford UK.

Malinen T. (2008). Estimating the long-run relationship between income inequality and economic development. Empirical Economics, Discussion paper, No. 634, 1–39 pp.

Malinen T. (2009). Estimating the long-run relationship between income inequality and economic development. Empirical Economics, Discussion paper, No. 260, 1–39 pp.

Malinen T. (2013). Inequality and growth: Another look with a new measure and method. Journal of International Development, Vol. 25(1), 122–138 pp.

Malinen T. (2012). Estimating the long-run relationship between income inequality and economic development. Empirical Economics, Vol. 42, 209–233 pp.

Martišius S. (2014). Statistikos metodai socialiniuose ekonominiuose tyrimuose. Vilniaus universiteto leidykla. ISBN 978-609-459-357-4.

Martišius S., Kėdaitis V. (2013). Statistika. Statistinės analizės teorija ir metodai. I dalis. Trečioji papildyta laida. Vilnius: Vilniaus universitetas.

Masten A. S. (2001). Ordinary Magic: Resilience Processes in Development. American Psychologist, Vol. 56, No. 3, 227–238 pp.

Masten A. S., Obradovic J., (2006). Competence and Resilience in Development. Annals of the New York Academy of Sciences, Vol. 1094, 13–27 pp.

Melnikas B., Baršauskas P., Kvainauskaite V. (2006). Transition Processes and Integral Cultural Space Development in Central and Eastern Europe: Main Problems and Priorities. Baltic Journal of Management, Vol. 1, No. 2.

Milanovic B. (2005). Half a World: Regional Inequality in Five Great Federations. Journal of the Asia Pacific Economy, Vol. 10, No. 4, 408–445 pp.

Milanovic B. (2010). Global Inequality Recalculated and Updated: the Effect of New PPP Estimates on Global Inequality and 2005 Estimates. The Journal of Economic Inequality, Springer; Society for the Study of Economic Inequality, Vol. 10, No. 1, 1–18 pp.

Milanovic B. (2014). The Return of 'Patrimonial Capitalism': A Review of Thomas Piketty's Capital in the Twenty–First Century. Journal of Economic Literature, Vol. 52, No. 2, 519–534 pp. Available online: https://www.aeaweb.org/articles?id=10.1257/jel.52.2.519.

Milanovic B. (2019). Capitalism, Alone: The Future of the System That Rules the World. Belknap Press, 304 p., ISBN 978-0674987593.

Misiūnas A. Bratčikovienė N. (2007). Pajamų nelygybė ir jų normalizavimas. Lietuvos statistikos darbai, Vol. 46, 112–135 pp.

MRU Quality of Life Laboratory Research (2016) publicated by Rakauskienė O. G., Puškorius S., Diržytė A., Servetkienė V., Krinickienė E., Bartuševičienė I., Volodzkienė L., Juršėnienė V. (2017). Socialinė ekonominė nelygybė Lietuvoje. Vilnius: Mykolo Romerio universitetas, 476 p. ISBN 978-9955-19-869-7.

Naujienų Centras (2019). TOP/500 turtingiausių žmonių Lietuvoje. 2019, Nr. 7. ISSN 2351-5864.

Nunez J., Tartakowsky A. (2007). Inequality of Outcomes vs. Inequality of Opportunities in a Developing Country. Estudios de Economia, Vol. 34, No. 2, 185–202 pp.

Nussbaum M., Sen A. (1993). The Quality of Life. Oxford, UK: Clarendon Press.

Organisation for Economic Co-operation and Development (OECD) (2004). Annual Report. 127 p. Available online: https://www.oecd-ilibrary.org/economics/oecd-annual-report-2004_annrep-2004-en.

Organisation for Economic Co-operation and Development (OECD) (2011). Divided We Stand: Why Inequality Keeps Rising. 400 p. Available online: https://www.oecd.org/els/soc/dividedwestandwhyinequalitykeepsrising.htm.

Okunevičiūtė Neverauskienė, L., Gruževskis, B. (2010). Lygybės statistikos pagrindai.// Nacionalinis lygybės statistikos veiksmų planas: mokslo darbai. Vilnius: Eugrimas.

Ostry J. D., Loungani P., Berg A. (2019). Confronting Inequality: How Societies Can Choose Inclusive Growth. Columbia University Press. ISBN 978-0-231-52761-3.

Pajuodienė G., Šileika A. (2001). Lietuvos gyventojų socialiniai sluoksniai (Viduriniosios klasės beieškant). Pinigų studijos, Vol. 3, 59–81 pp.

Park H. M. (2011). Practical Guides to Panel Data Modeling: A Step by Step Analysis Using Stata. International University of Japan, 52 p.

Perrotti R. (1996). Growth, Income Distribution, and Democracy: What the Data Say. Journal of Economic Growth, Vol. 1, No. 2, 149–187 pp.

Persson, T., Tabellini, G. (1994). Is Inequality Harmful for Growth? American Economic Review, Vol. 84, No. 3, 600–621 pp.

Persson T., Tabellini G. (1991). Is Inequality Harmful for Growth? Theory and Evidence. Working Paper, No. 91–155, 1–37 pp.

Piketty T. (2014), Capital in the Twenty-First Century, Harvard University Press, 685 p., ISBN 978-0-674-43000-6.

Piketty T. (2015). The Economics of Inequality. Belknap Press, 160 p., ISBN 978-0674504806.

Piketty T. (2016). Chronicles: On Our Troubled Times. Printed in Great Britain by Clays Ltd, St Ives Plc, 181 p., ISBN 978-0-241-23491-4.

Piketty T. (2020). Capital and Ideology. Belknap Press, 1104 p., ISBN 978-0674980822.

Piketty T., Saez E. (2003). Income Inequality in the United States, 1913–1998. Quarterly Journal of Economics, Vol. 118, 1–39 pp.

Piketty T., Saez E. (2006). The Evolution of Top Incomes: A Historical and International Perspective. American Economic Review: Papers and Proceedings, Vol. 96.

Piketty T., Zucman, G. (2015). Wealth and Inheritance in the Long Run. Handbook of Income Distribution. Elsevier B.V., Vol. 2B. ISSN 1574-0056. Available online: http://piketty.pse.ens.fr/files/PikettyZucman2014HID.pdf.

Pilling D. (2018). The Growth Delusion: The Wealth and Well-Being of Nations. Bloomsbury Publishing Plc, 338 p., ISBN 978-1-4088-9370-8.

Puškorius S. (2016). Pajamų ir Vartojimo Pasiskirstymo Lietuvoje Analizė. Viešoji Politika ir Administravimas. T. 15, Nr. 4, 608–622 pp.

Quinn A. M. (2006). Relative Deprivation, wage differentials and mexican migration. Review of Development Economics, Vol.10, Issue 1, 135–153 pp.

Rakauskienė O. G. (2010). Lietuvos gyventojų gyvenimo gerovės raida ir perspektyvos. Regnum est. 1990 m. Kovo 11-osios Nepriklausomybės Aktui – 20. Liber Amicorum Vytautui Landsbergiui: mokslo straipsnių rinkinys. Vilnius: Mykolo Romerio universiteto Leidybos centras.

Rakauskienė O. G., Lisauskaitė V. (2009). Quality of Life of the Population of Lithuania: The Conception, Development and Prospects. Ekonomika. Mokslo darbai, T. 88.

Rakauskienė O. G., Servetkienė V., Puškorius S., Čaplinskienė M., Diržytė A., Ranceva O., Bilevičienė T., Kazlauskienė E., Žitkienė R., Štreimikienė D., Monkevičius A., Bieliauskienė R., Laurinavičius A., Krinickienė E. (2015). Gyvenimo kokybės matavimo rodiklių sistema ir vertinimo modelis. Vilnius: Mykolo Romerio universitetas, 760 p., ISBN 978-19-713-3.

Rakauskienė O. G., Puškorius S., Diržytė A., Servetkienė V., Krinickienė E., Bartuševičienė I., Volodzkienė L., Juršėnienė V. (2017). Socialinė ekonominė nelygybė Lietuvoje. Vilnius: Mykolo Romerio universitetas, 476 p., ISBN 978-9955-19-869-7.

Rakauskienė, O. G., Servetkienė, V. (2011). Lietuvos gyventojų gyvenimo kokybė: dvidešimt metų rinkos ekonomikoje. Vilnius: Mykolo Romerio universitetas, 360 p., ISBN 978-9955-19-317-3.

Rawls J. (1971). A Theory of Justice. Cambridge, Massachusetts: The Belknap Press of Harvard University Press, 607 p. ISBN 0674880145 9780674880146.

Reich R. (2010). Aftershock: The Next Economy and America's Future. Knopf, 192 p., ISBN 978-0307592811.

Reich R. (2012). Beyond Outrage: Expanded Edition: What Has Gone Wrong with Our Economy and Our Democracy, and How to Fix It. Vintage, 176 p., ISBN 978-0345804372.

Reich R. (2016). Saving Capitalism: For the Many, Not The Few. Icon Books Ltd, 279 p., ISBN 978-178578-176-6.

Reich R. (2017). Economics in Wonderland: Robert Reich's Cartoon Guide to a Political World. Fantagraphics Books, 117 p., ISBN 9781683960607.

Reich R. (2019). The Common Good. Vintage, 208 p., ISBN 978-0525436379

Reinhardt C., Rogoff K. S. (2011). This Time Is Different: Eight Centuries of Financial Folly. Princeton University Press, 512 p., ISBN 978-0691152646.

Rogers L. O. (2019). Commentary on Economic Inequality: "What" and "Who" Constitutes Research on Social Inequality in Developmental Science? Developmental Psychology, Vol. 55, No. 3, 586–591 pp.

Rothbard M. N. (2016). Anatomy of the State. Martino Fine Books, 30 p. ISBN 978-1614279884.

Rothbard M. N. (2006). For a New Liberty: The Libertarian Manifesto. Ludwig von Mises Institute, Auburn, Alabama, 420 p. ISBN 978-0-945466-47-5.

Rudzkienė V. (2005). Socialinė statistika. Vilnius: Mykolo Romerio universiteto leidybos centras.

Saez E., Zucman G. (2016). Wealth Inequality in the United States since 1913: Evidence from Capitalized Income Tax Data. The Quarterly Journal of Economics, Vol. 131, No. 2, 519–578 pp.

Saez E., Zucman G. (2019). The Triumph of Injustice. How the Rich Dodge Taxes and How to Make Them Pay. W. W. Norton & Company, 232 p., ISBN 978-1324002727.

Salverda W., Nolan B., Smeeding T. M. (2013). The Oxford Handbook of Economic Inequality. Oxford University Press, 736 p., ISBN 978-0-19-960606-1.

Seimas of the Republic of Lithuania (2019). LR 2020 metų valstybės biudžeto ir savivaldybių biudžetų finansinių rodiklių patvirtinimo įstatymas. 2019 m. gruodžio 17 d. Nr. XIII-2695.

Sen A. (1997). From Income Inequality to Economic Inequality. Southern Economic Journal, Vol. 64, No. 2, 384–401 pp.

Sen A. (2004). Rationality and Freedom. Belknap Press, 752 p., ISBN 978-0674013513.

Sen A. (2015). The economic consequences of austerity. New Statesman. Available online: https://www.newstatesman.com/politics/2015/06/amartya-sen-economic-consequences-austerity.

Sen A. (1999). The Possibility of Social Choice. American Economic Review, 89 (3), 349–378 pp.

Sen A. (1997). On economic inequality. Oxford university press, 276 p. ISBN 9780198292975.

Sen A. (1993). Capability and Wellbeing. In M. Nussbaum, & A. Sen (Eds.). The Quality of Life, Oxford: Clarendon Press, 30–53 pp.

Servetkienė V. (2013) Gyvenimo kokybės daugiadimensis vertinimas, identifikuojant kritines sritis. Daktaro disertacija. Mykolo Romerio Universitetas.

Sierminska E., Brandolini A., Smeeding T. M. (2006). The Luxembourg Wealth Study – A cross-country comparable database for household wealth research. The Journal of Economic Inequality, Vol. 4, 375–383 pp.

Šileika A. (2010). Aktualūs investicijų į Žmogiškąjį kapitalą ir pajamų paskirstymo bei perskirstymo klausimai. Ekonomika ir vadyba: aktualijos ir perspektyvos, Vol 1, No. 17. 147–149 pp.

Šileika A., Tamašauskienė Z., Zaleskis E. (2009). Gyvenimo lygis ir jo lyginamoji analizė Lietuvoje ir kitose Europos Sąjungos Šalyse. Socialiniai tyrimai, Vol. 3, No. 17. Šiauliai: Šiaulių universiteto leidykla, 84–95 pp. ISSN 1392-3110.

Šileika A., Zabarauskaitė R. (2009). Skurdas: metodologijos klausimai ir lygis Lietuvoje. Socialinis darbas. Mokslo darbai, Vol. 8, No. 1. Vilnius: Mykolo Romerio universitetas, 17–27 pp. ISSN 1648-4789.

Sim S. (2017). Insatiable: The Rise and Rise of the Greedocracy. Reaktion Books Ltd., 192 p., ISBN 978-1-78023-734-3.

Skučienė D. (2008). Pajamų nelygybė Lietuvoje. Filosofija. Sociologija, Vol. 19, No. 4, 22–33 pp.

Solow R. M. (1986). On the Intergenerational Allocation of Natural Resources. Scandinavian Journal of Economics, Vol. 88, No. 1, 141–149 pp.

Solow, R. M. (1956). A Contribution to the Theory of Economic Growth. The Quarterly Journal of Economics, Vol. 70, No. 1, 65–94 pp.

Solow, R. M. (1974). The Economics of Resources or the Resources of Economics. The American Economic Review, Vol. 64, No. 2, 1–14 pp.

Starkauskienė V. (2011). Gyvenimo kokybės veiksniai ir jos kompleksinio vertinimo modelis. Daktaro disertacija, Kaunas: Vytauto Didžiojo Universitetas.

Stiglitz J. E. (2009). Towards a better measure of well-being. The Financial Times. Available online: https://www.ft.com/content/95b492a8-a095-11de-b9ef-00144feabdc0.

Stiglitz J. E. (2010). Freefall: Free Markets and the Sinking of the Global Economy. Penguin Books, 443 p., ISBN 978-0-141-04512-2.

Stiglitz J. E. (2010). The Stiglitz Report: Reforming the International Monetary and Financial Systems in the Wake of the Global Crisis. The New Press, 240 p., ISBN 978-1595585202.

Stiglitz J. E. (2012). The Price of Inequality: How Today's Divided Society Endangers Our Future. W. W. Norton & Company, 560 p., ISBN 0393345068.

Stiglitz J. E. (2014). Reforming Taxation to Promote Growth and Equity. Roosevelt Institute. Available online: https://rooseveltinstitute.org/wp-content/uploads/2020/07/RI_Reforming_Taxation_White_Paper_201405.pdf.

Stiglitz J. E. (2015). Rewriting the Rules of the American Economy. An Agenda for Growth and Shared Prosperity. W. W. Norton & Company, 256 p., ISBN 978-0-393-35312-9.

Stiglitz J. E. (2015). The Great Divide. Penguin Books, 438 p., ISBN 978-0-141-98122-2.

Stiglitz J. E. (2016). How to Restore Equitible and Sustainable Economic Growth in the United States. American Economic Review: Papers & Proceedings, Vol. 106, No. 5, 43–47 pp.

Stiglitz J. E. (2016). The Euro: And Its Threat to the Future of Europe. Published in Great Britain by Allen Lane, 459 p., ISBN 978-0-141-98324-0.

Stiglitz J. E. (2017). Globalization and Its Discontents Revisited: Anti-Globalization in the Era of Trump. Penguin Books, 472 p., ISBN 978-0-141-98666-1.

Stiglitz J. E. (2018). Inequality and Economic Growth. Roosevelt Institute. Available online: https://rooseveltinstitute.org/wp-content/uploads/2020/07/RI-Inequality-and-Economic-Growth-201803.pdf.

Stiglitz J. E. (2020). People, Power, and Profits: Progressive Capitalism for an Age of Discontent. W. W. Norton & Company, 400 p., ISBN 978-0393358339.

Stiglitz J. E., Sen A., Fitoussi J. P. (2010). Report by the Commission on the Measurement of Economic Performance and Social Progress. Paris: Commission on the Measurement of Economic Performance and Social Progress. Available online: http://www.stiglitz-sen-fitoussi.fr/documents/rapport_anglais.pdf.

Stilwell F. (2016). Why Emphasise Economic Inequality in Development?. Journal of Australian Political Economy, No. 78, 24–47 pp.

Stirati A. (2017). Wealth, Capital and the Theory of Distribution: Some Implications for Piketty's Analysis. Review of Political Economy, Vol. 29, No. 1, 47–63 pp.

Storm S., Naastepad C. W. M. (2015). Europe's Hunger Games: Income Distribution, Cost Competitiveness and Crisis. Cambridge Journal of Economics, Vol. 39, No. 3, 959–986 pp.

Tamašauskienė, Z., Šeputienė, J., Balvočiūtė, R., Beržinskienė-Juozainienė, D. (2016). Darbo pajamų dalies kitimo poveikis bendrajai paklausai. Mokslo studija. Šiauliai: Šiaulių universitetas.

Tarptautinis viešosios, socialinės ir kooperatinės ekonomikos mokslinių tyrimų ir informacijos centras (CIRIEC) (2007). Europos Sąjungos socialinė ekonomika. Europos ekonomikos ir socialinių reikalų komitetui skirto pranešimo santrauka. Europos ekonomikos ir socialinių reikalų komitetas, Vol. 2007, 26 p., ISBN 978-92-830-0866-8.

Telešienė A. (2017). Ekonomikos augimas nesustabdys emigracijos, kol išliks socialinė nelygybė. Mokslo Lietuva. Available online: http://mokslolietuva.lt/2017/12/a-telesiene-ekonomikos-augimas-nesustabdys-emigracijos-kol-isliks-socialine-nelygybe/.

Tridico P., Pariboni R. (2018). Inequality, Financialization, and Economic Decline. Journal of Post Keynesian Economics, Vol. 41, No. 2, 236–259 pp.

United Nations Department of Economic and Social Affairs (2017). Combating Inequalities to End Poverty. United Nations Headquarters, New York.

Volodzkienė L. (2020) Ekonominės nelygybės poveikio socialinei ekonominei pažangai vertinimas. Daktaro disertacija. Mykolo Romerio Universitetas.

Wilkinson R., Pickett K. (2018). The Inner Level: How More Equal Societies Reduce Stress, Restore Sanity and Improve Everyone's Well-Being. London: Allen Lane, 352 p. ISBN 978-1846147418.

Zabarauskaitė R., Blažienė I. (2012). Gyventojų pajamų nelygybė ekonominių ciklų kontekste. Verslas: Teorija ir praktika, Vol. 13, No. 2, 107–115 pp.

Index

Printed in the United States
by Baker & Taylor Publisher Services

Printed in the United States
by Baker & Taylor Publisher Services